Occupational Health and Safety in Ontario

Nora Rock

2008
Emond Montgomery Publications Limited
Toronto, Canada

Copyright © 2008 Emond Montgomery Publications Limited. All rights reserved. No part of this publication may be reproduced, stored in a retrieval system, or transmitted, in any form or by any means, photocopying, electronic, mechanical, recording, or otherwise, without the prior written permission of the copyright holder.

Emond Montgomery Publications Limited
60 Shaftesbury Avenue
Toronto ON M4T 1A3
http://www.emp.ca/college

Printed in Canada.

We acknowledge the financial support of the Government of Canada through the Book Publishing Industry Development Program (BPIDP) for our publishing activities.

Acquisitions and developmental editor: Peggy Buchan
Marketing manager: Christine Davidson
Copy editor: Sarah Gleadow
Production editor: Jim Lyons, WordsWorth Communications
Proofreader and indexer: Paula Pike, WordsWorth Communications
Typesetter: Debbie Gervais, WordsWorth Communications
Cover designer: John Vegter

Library and Archives Canada Cataloguing in Publication

Rock, Nora, 1968-
    Occupational health and safety in Ontario / Nora Rock.

Includes index.
ISBN 978-1-55239-272-0

    1. Industrial hygiene—Law and legislation—Ontario. 2. Industrial safety—Law and legislation—Ontario. 3. Industrial hygiene—Ontario—Textbooks. 4. Industrial safety—Ontario—Textbooks. I. Title.

KEO673.R63 2008   344.71304'65   C2008-901431-6
KF3570.R63 2008

Contents

CHAPTER 1 Thinking About Safety at Work

Introduction	1
Perspectives on Safety at Work	2
Employee/Union Perspective	2
Employer/Management Perspective	4
Society's Perspective	6
Government's Perspective	7
Approaches to Workplace Safety	8
Key Terms	9
Review Questions	10
Discussion Questions	10

CHAPTER 2 The Evolution of Workplace Safety Law

Introduction	11
Compensation for Injured Workers	12
In Other Jurisdictions	12
In Canada	13
Legislating Workplace Safety	15
Introduction	15
The Constitutional Context of Safety Legislation	16
Environmental Protection and Workplace Safety	17
Introduction	17
Overview of Environmental Law in Canada	19
Key Terms	21
Review Questions	21
Discussion Questions	22

CHAPTER 3 Overview of the OHSA

Introduction	23
Structure of the OHSA	24
Table of Contents	24
Definitions	24
Application	25
Administration	26
Duties of Employers and Other Persons	27
Codes of Practice	27
Toxic Substances	28
Right to Refuse or Stop Work Where Health or Safety in Danger	28
Reprisals by Employer Prohibited	30
Notices	30
Enforcement	30
Offences and Penalties Under the OHSA	32
Regulations	32
The Internal Responsibility System	34
Introduction	34
Employers	35
Supervisors	38
Health and Safety Representative or Joint Health and Safety Committee	39
Trade Unions	39
Workers	42
Health and Safety Representatives and Joint Health and Safety Committees	44
Introduction	44
Health and Safety Representatives	44
Joint Health and Safety Committees	45
Mandate and Duties	47
Procedures	48
Key Terms	50
Review Questions	50
At Issue: Is Total Injury Prevention a Realistic Goal?	51

CHAPTER 4 Introduction to the WHMIS

Introduction	53
Overview of the System	54
History	54
Legislative Framework	55
Classification of Hazardous Materials	57
Supplier and Importer Compliance Requirements	57
Introduction	57
Classification	58
Disclosure by Suppliers and Importers	60
WHMIS Compliance for Employers	66
Introduction	66
Identification/Classification of Controlled Products	66

Labels and MSDSs	67
Worker Training	68
Key Terms	69
Review Questions	69

CHAPTER 5 Compensation for Injured Workers

Introduction	71
Overview of the Legislation	72
Application	72
Who Is a "Worker"?	75
How the Fund Works	75
Employer Duties Under the WSIA	78
Understanding and Managing Premiums	79
Understanding Benefits	84
Managing Workers' Return to Work	85
Managing Fraud and Suspected Fraud in the WSIA System	85
Worker Fraud and Non-Compliance	85
Employer Fraud and Non-Compliance	88
Key Terms	90
Review Questions	90
At Issue: Should the WSIB provide compensation for chronic pain?	91

CHAPTER 6 Other Workplace Safety Laws

Introduction	93
The Industrial Establishments Regulation	94
Introduction	94
Pre-Start Health and Safety Reviews	96
Legislated Technical Standards	98
Introduction	98
Overview of Legislated Technical Standards	98
Voluntary Standards	104
Introduction	104
Sources of Industrial Safety Standards	105
Subject-Specific Laws with Safety Aspects	106
Human Rights Law	108
Introduction	108
Safety-Related Human Rights Issues	109
Ontarians with Disabilities Act	113
The Criminal Law	113
Introduction	113
Workplace Violence	114
Criminal Negligence	116
Key Terms	118
Review Questions	118

CHAPTER 7 OHS from a Strategic Perspective

Introduction	119
Should Occupational Health and Safety Be Subject to a Cost–Benefit Analysis?	121
The Cost of Workplace Injuries	123
Introduction	123
Property Damage	123
Lost Productivity	124
Recruitment and Training Costs	124
Administrative Costs	125
Fines and Other Penalties	125
Impact on Public Image	127
Costs to Employees, Families, and the Community	127
Compliance and Prevention Costs	129
Categories of Costs	129
Justifying OHS Costs	130
Opportunities to Offset OHS Costs	131
Measuring Results	132
Safety Leadership	134
Key Terms	136
Review Questions	136

CHAPTER 8 Steps to Compliance I

Introduction	137
Setting Goals	138
Introduction	138
The Role of Senior Management	138
Building a System	139
CSA-Z1000: A Blueprint for a Health and Safety System	140
Assessing the Status Quo	142
Introduction	142
Reviewing Injury and Accident Statistics	143
Reviewing Existing Policies and Procedures	144
Evaluating Culture, Perceptions, and Behaviours	145
Evaluating Formal Compliance and Reporting	146
Planning for Change	146
Marshalling Resources	147
Implementing Changes	150
Introduction	150
Health and Safety Representative or JHSC	150
WHMIS Compliance	150
Scheduling Inspections	151
Developing and Renewing Policies and Procedures	151
Managing Training	153

Introducing Health Promotion Programs .. 155
Managing Investigations 157
Motivating, Rewarding, and Evaluating Progress .. 157
Introduction 157
Motivation 157
Evaluation 158
Setting New Goals 159
Review Questions 159

CHAPTER 9 Steps to Compliance II (Post-Incident)

Introduction 161
What Is an Incident? 162
Why Track Environmental and Human Rights Incidents? 164
Basic Steps Following an Incident 166
Complaints About Hazards/Conditions 166
Internal Complaints 167
External Complaints 168
Work Refusals or Stoppages 169
Accidents Without Injuries 171
First Aid/No-Lost-Time Injuries 171
Lost-Time/Medical-Attention Injuries 172
Immediate Considerations 172
Reporting Obligations Generally 173
Reporting Obligations Under the OHSA ... 174
Reporting Obligations Under the WSIA 174
Critical Injuries and Fatalities 177
Occupational Illnesses 179
What Is an Occupational Illness? 179
What to Do When a Worker Reports an Occupational Illness 179
Traumatic Events 180
Investigating Incidents 182
Introduction 182
The Investigation Team 182
Investigation Objectives 183
The Importance of Follow-up Actions 183
Key Terms 184
Review Questions 184
At Issue: Is the Investigation of Work Refusals an Essential Service? 185

CHAPTER 10 Facing an Investigation or Prosecution

Introduction 187
Ministry of Labour Inspections 188
Why They Happen 188
What to Do 189
Inspectors' Orders 190
Appealing MOL Orders 191
Legal Representation 193
Ministry of Labour Investigations and Prosecutions 194
Inspections and Investigations: What's the Difference? 194
What to Do 196
Preparing for and Attending Court 199
Factors Influencing Fines 201
Minimizing Bad Press 201
WSIB Investigations and Prosecutions 202
Workwell Audits 202
Grievances in Unionized Workplaces 203
Criminal Prosecutions 204
Coroners' Inquests 206
Introduction 206
What Triggers a Coroner's Investigation of a Workplace Death? 206
What Is the Role of the Employer? 207
Key Terms 207
Review Questions 208

CHAPTER 11 Managing WSIA Claims and the Transition Back to Work

Introduction 209
Legal Obligation to Promote Return to Work 210
Introduction 210
Assessing Return Date, Abilities, and Limitations 210
Identifying Safe and Suitable Work 213
Accommodating the Worker 214
When Does the Duty to Support ESRTW Expire? 216
The Worker's Obligations 217
Strategies for Managing the Return to Work 218
Introduction 218
Post-Accident Communication 218
Planning for Accommodation 220
Timing and Preparations 221
Preparing Co-Workers 221
Complaints and Problems in the Return-to-Work Process 222
What Happens When a Worker Is Unlikely to Return? 224
Review Questions 225

CHAPTER 12 A Culture of Safety

Introduction 227
What is a Culture of Safety, and Why Strive for One? 228

What Does a Culture of Safety Look Like?	229	**Appendix D**	
Assessing the Status Quo	229	Sample Material Data Safety Sheet (MSDS) for	
Fostering Safety Beliefs	231	Silica Sand	247
Introduction	231	**Appendix E**	
Risk Adversity, and Promoting Risk		Sample Form for Use in Creating a Material	
Intolerance	231	Safety Data Sheet	251
Building Worker Pride in the Organization's		**Appendix F**	
Safety Record	233	Costing an Individual Incident/Injury Form—	
Promoting Safe Behaviours	233	from the WSIB	255
Introduction	233	**Appendix G**	
Information and Training	234	Resources for Safety Managers, Health and Safety	
Reminders	234	Representatives, and JHSC Members	259
Feedback	236	**Appendix H**	
Rewards	236	Workplace-Specific Hazard Training	
Key Terms	237	Information Form	263
Review Questions	238	**Appendix I**	
Appendix A		WSIB Form 0007A—Employer's Report	
Regulations Under the Ontario Occupational		of Injury/Disease	265
Health and Safety Act	239	**Appendix J**	
Appendix B		Personal Perceptions of Organizational Safety	269
General and Sector-Specific Health and Safety			
Associations	241	References	271
Appendix C		Glossary	275
Guidelines for the Use of Professional Judgment		Index	279
in Classifying Controlled Products	243		

CHAPTER 1

Thinking About Safety at Work

CHAPTER OBJECTIVES

After completing this chapter, you should be able to:

- Understand that different stakeholders have different perspectives on workplace safety.
- Explain why employees care about workplace safety.
- Describe some of the challenges employers face in promoting workplace safety.
- List several ways in which workplace accidents and illnesses affect the broader community and society as a whole.
- Describe the ways in which workplace injuries and accidents consume government resources.
- Understand why a multidisciplinary approach to safety in the workplace holds the greatest promise for workplace injury and illness prevention.

INTRODUCTION

This is a book about the law.

Because work is a central feature of most people's lives, workers, employers, lawmakers, and human rights advocates have always been motivated to take steps to ensure that the hours people spend at work are not only productive, but also comfortable. Since the Industrial Revolution, the safety of workers has been a key workplace issue. Today, workplace safety in Ontario is governed by a well-coordinated set of legal schemes that address many facets of work safety: hazardous materials; employer, supervisor, and worker safety responsibilities; and a public insurance scheme to address work injury **claims**, among others. Ontario employers are required to maintain a current understanding of all of these schemes so that they can run their businesses in compliance with the law. This book is designed to support the development of that understanding.

claim
an application for compensation under the terms of an insurance policy

But let's take a step back, for a moment, and consider the issue of workplace safety from a broader perspective.

First, whose interests are served by a commitment to workplace safety? "Workers and their families" is promoting the obvious answer. However, even a few moments' thought about the issue should convince you that allowing people to be hurt or sickened at work affects business, too. Good managers understand that good people are a company's greatest asset. Considerable resources are invested into each new hire, and if an employee is lost due to injury or illness, not only will productivity be affected but that investment may be placed in jeopardy, too.

There are also broader labour market implications of workplace injuries. If a particular industry develops a reputation as being dangerous, employers in that industry will face recruitment challenges, and wage demands may rise in response to candidates' perceptions of risk.

On a community level, serious workplace injuries can lead to greater demands on limited social service resources. Workplace deaths can devastate communities and reduce tolerance for the presence, in the community, of industries that are perceived as dangerous, dirty, or polluting. Reduced community tolerance for a particular industry can affect that industry's economic growth.

To help you understand the importance of promoting safety in the place in which you work, the following section will introduce a range of perspectives on the issue: the perspective of unions and workers; the perspective of the employer/management; society's perspective; and the perspective of government. The next section will then discuss, in general terms, approaches to safety promotion that have been developed by the disciplines of law, engineering, and psychology/sociology.

With this context in place, the rest of this book will focus more narrowly on legal approaches to workplace safety and, more specifically, on employers' compliance responsibilities.

PERSPECTIVES ON SAFETY AT WORK

Employee/Union Perspective

INJURY AND ILLNESS AVOIDANCE

Unless a job has a reputation for carrying with it a particularly high degree of risk, the possibility of being injured, sickened, or disabled at work is generally not prominent in the minds of new workers. In fact, some employees find the safety policies and procedures implemented by their employers to be downright bothersome. For example, employees who are required to wear personal protective equipment (such as safety eyewear) may complain about physical discomfort. Employees who must comply with lockout procedures before adjusting equipment may feel that these procedures slow down their work and reduce productivity.

However, workers' perception of the risks involved in their work can shift quickly in response to experience. Experiences that tend to highlight risk from an employee's perspective can include:

- a near-hit incident, in which the employee narrowly escapes injury;
- a minor accident that could have been much worse;

- experiencing discomfort (even if no health consequences result) from exposure to workplace materials (for example, a solvent that emits strong fumes); and

- witnessing or hearing about an injury to a co-worker.

Psychologists who study workplace safety tend to believe that this shift in risk tolerance is actually beneficial in helping to reduce accidents because it promotes caution on the part of workers. In Ontario in recent years, a media campaign targeted specifically at younger workers has been in place. The campaign was created in response to findings that young workers have a statistically high rate of workplace injury and a generally inadequate perception of workplace risks.

Once employees become aware of the risks involved in their work, they often demand adequate protection from those risks. The reaction of the employer to these demands has a significant effect on the safety climate (or lack thereof) that will emerge in the workplace. The same is true of the reaction of the employer to actual workplace accidents, because accidents that happen to others often have a significant impact on the injured person's colleagues.

As you will learn later in this chapter, safety initiatives often cost employers money, and it can be difficult to measure how—or even whether—promoting safety has a positive impact on productivity. For this reason, there is tension in some workplaces between employees who would like to stay safe and employers who would like to limit the costs of doing business.

In unionized workplaces, worker **unions** often play a key role in encouraging management to take safety seriously. Unions can provide information about risks and risk-reduction strategies, they can mediate worker–management conflicts over safety, they can negotiate the introduction of new safety initiatives, and they can provide support to injured workers in dealing with management and with outside agencies. For an example of a successful union-driven initiative to change occupational health and safety law, see the Success Story feature below.

union
an association of workers formed to protect and advance their rights and interests

☀ Success Story

UFCW Union Leads Campaign to Extend OHSA Protection to Agricultural Workers

On June 30, 2006 amendments to the Ontario *Occupational Health and Safety Act* **(OHSA)** extended the legislation's protection to agricultural workers.

Before the introduction of these amendments, Ontario was the only province to exclude agricultural workers from the application of the OHSA, though Alberta and Prince Edward Island have restrictions on the application of the Act to that sector.

The change to the law was influenced in large part by a three-year campaign led by the United Food and Commercial Workers (UFCW) Canada union. The campaign, which included both media awareness-raising activities and legal action, sought to highlight the UFCW's position that "the exclusion of these workers from OHSA protection was unfair, unjust and a violation of their Charter rights."

The legislative changes gave Ontario farm workers the right, as well as the obligation, to form workplace health and safety committees and to develop a farm safety plan and policy in conjunction with their employer. It also gave them the right to be told about workplace hazards and dangers, such as toxic chemicals, and the right to

Occupational Health and Safety Act **(OHSA)**
an Ontario Statute that regulates the promotion of worker health and safety in most Ontario workplaces

refuse work that was dangerous or unsafe without having to worry about being punished by their employer.

Prior to these changes, the UFCW provided counselling and training about occupational health and safety for farm workers. In this capacity, the union prepared an occupational health and safety manual in English and Spanish that it distributed to migrant farm workers' resource centres in Ontario and Quebec.

While employers may have honest concerns about safety in their workplaces, for employees safety is an acutely important issue. For an employee, unsafe work carries with it the possibility of injury, disability, and the inability to work at all or to provide for dependants. While to employers injuries mean lost people-hours and reduced productivity, to workers injuries can mean pain, loss of income and key relationships, mental health changes, and a (sometimes permanent) reduction in their ability to enjoy the pleasures of life.

FAIR COMPENSATION

When an employee is injured or sickened, his or her focus shifts from avoidance to compensation and accommodation.

Many injured employees suffer real economic losses as a result of their injuries. To some degree, **workers' compensation** programs have emerged to offset some of these losses. It is in the interests of workers to have an adequately funded compensation system in place and to have access to publicly funded health care. However, there are some costs that are not covered by these systems, such as post-injury trauma counselling and non-OHIP health services. Employers are required to continue to pay **benefits** to workers who are off the job after an injury, and in some cases employee benefits offset an additional portion of post-injury costs. Workplaces that provide additional support to injured workers—for example, support groups where injured workers can talk about their experiences—go further toward supporting employees in their recovery.

workers' compensation
money paid to a person for an injury suffered on the job

benefits
money to which a person is entitled from a pension plan, government support program, etc.

When an employee is ready to return to work after suffering a disabling injury, the employee is entitled to have the employer accommodate, to the extent reasonably possible, the employee's new range of capabilities. For employees, accommodation can make the difference between being able to work again or not. Employees who are actively accommodated and supported upon their return to work are most likely to be able to continue to be productive for the organization; this continued productivity, in turn, protects the employer's past recruitment and training investment in the employee.

Unions can assist employees to recover financially and physically from workplace injuries by negotiating optimum employee benefits packages and special supports for injured workers, and by encouraging employers to actively accommodate returning workers.

Employer/Management Perspective

IMPACT OF WORKPLACE INJURIES AND ILLNESS

Most management staff sincerely care about their employees on a personal level. The prospect of having a worker contract a chronic illness, become permanently disabled, or die on the job weighs heavily on the minds of most managers, particularly those who have witnessed serious accidents in the past.

From a business perspective, as manufacturing and other workplaces have become more automated, workforces have been downscaled; the remaining workers are, on average, more skilled. Employers have typically invested a considerable amount of training and mentoring in these employees to prepare them for the work they do. In non-industrial workplaces, employees may have, through experience, developed an intimate knowledge of the company's strengths, weaknesses, and critical needs. Rather than in its computers, a company's organizational memory is contained, in large part, in its employees' brains; the organization's corporate culture and vision is a product of the employees' individual personalities working together. Losing even one key employee to injury or illness, then, can be a serious business disadvantage in any kind of business. It is easy to understand how an organization's heart and soul is in its people.

From a public relations perspective, a serious workplace accident can directly affect a company on multiple levels. Having a reputation as a shop where people get hurt or killed on the job can make it difficult for the company to recruit the necessary labour. Employees may also request higher pay in return for shouldering the risk of harm. From a sales perspective—particularly in smaller communities—being known as a company that has caused a death or serious disability can directly affect sales.

Within the company, an injury to a co-worker can affect the morale of the rest of the staff. After a serious injury, the way in which the employer reacts to the incident has significant implications from a morale perspective. Employers who are seen to be supportive of the injured worker's recovery and who take clear steps to eliminate the hazard that caused the injury, if any, are the most likely to avoid enduring morale problems. (Also, an employer's reaction—or lack thereof—to a serious workplace incident can influence the fine that may be imposed on the employer following a prosecution.)

ECONOMICS OF WORKPLACE SAFETY

Most employers now view running a safe workplace as both a legal and a moral obligation. However, many safety-promotion measures come with a price tag. The costs of safety compliance go beyond the purely monetary: not only must employers purchase and maintain safety equipment and supplies, they must also devote personnel resources to safety training, inspections, accident investigation, and other safety-related tasks. Finally, some safety procedures (for example, rest breaks for employees working in uncomfortable conditions) can slow down the pace of work, which affects productivity (Geller, 1996).

For start-up or smaller companies in particular, safety compliance represents a considerable expense. In addition, the benefits of safety measures are hard to quantify—after all, it's impossible to count avoided accidents. As you will learn in chapter 7, which examines occupational health and safety from a strategic perspective, applying a cost–benefit analysis to safety compliance is problematic for a number of reasons. Businesspeople are taught to make decisions about expenditures based on realistic predictions of the potential of those expenditures to contribute to corporate profits. In the case of workplace safety expenditures, however, those predictions cannot be made with any degree of accuracy. Often, safety-promotion initiatives must simply be adopted "on faith," and this attitude runs counter to conventional business thinking.

One fact, however, is clear: workplace injuries and illnesses are unpredictable and expensive. As you will see in the next section (and if you try out the Injury Cost Calculator mentioned in the Thinking Safe feature, below), the potential costs to an employer of a single serious accident are staggering. The case for managing the risk of workplace accidents is compelling from both a moral and an economic perspective. For employers, the biggest challenge when it comes to occupational health and safety is to identify and implement safety solutions that manage the risk of harm to employees without threatening the financial survival of the business.

> ## Thinking Safe
> ### Injury Cost Calculator
> The Ontario Safety Association for Community and Healthcare (OSACH), a highly respected sector-specific safety association, has created an "Injury Cost Calculator" designed to demonstrate the cost of a single workplace injury to an employer. This fascinating interactive tool is available at www.osach.ca/misc_pdf/InjuryCostCalculator.

Society's Perspective

The full impact of workplace injuries on communities and on society as a whole is not yet well understood. While we know at an intuitive level that injured workers struggle with their other roles in life (for example, keeping house, parenting, enjoying hobbies and recreation, participating in volunteerism and community projects), these impacts have rarely been formally quantified.

However, the issue is attracting increasing attention. In 1999 the (US) National Institute for Occupational Safety and Health (NIOSH), a division of the Centers for Disease Control and Prevention (CDC), held a major conference titled *Functional, Economic, and Social Outcomes of Occupational Injuries and Illnesses Conference: Integrating Social, Economic, and Health Services Research*. The objective of this conference was to set the stage for collecting and analyzing data about the impact of workplace injuries at the level of communities and society.

In New Zealand, a 2002 government report entitled *Aftermath* (prepared with the purpose of presenting the results of a research study called *The Social and Economic Consequences of Workplace Injury and Illness Study*) analyzed a number of individual accounts of workplace injuries and deaths, and emphasized the responsibility of the community as a whole to reduce its tolerance for workplace accidents:

> We are all, one way or other, directly or indirectly, responsible for the prevention of harm at work or for the care of those harmed. This is a community issue and requires all those involved in workplace health and safety (workers, employers, their families and government) to approach each other with a community of interest in better prevention and care. (Adams et al., 2002)

The same study describes the "weight of suffering and loss" to the community from occupational accidents as "a drag on growth, a brake on success and happiness," and concludes that if you "[m]ultiply the stories literally hundreds and thousands of times ... you can begin to understand the level of waste, suffering and loss that unnecessary occupational illness and injury produces."

The study found effects from serious workplace injuries in five areas: the individual, the individual's family and friends, the individual's workplace, the medical sector, and the government.

On a family level, the study found that "[r]elationships with family and friends were strained through emotional stress, financial pressure or physical isolation. Family and friends were deeply impacted, leading to either deeper bonding or disintegration of the relationship. In a number of cases the injury or illness resulted in temporary or permanent loss of sexual or other intimacy between partners." In some cases, injured workers became suicidal, depressed, or violent.

The study also found disruptions to the parent–child relationship, with alienation of children from injured parents and necessary changes to caregiving arrangements in some cases. In other cases, the injury resulted in limitations on the educational and career prospects of the injured worker's spouse.

On a community level, injuries to workers often lead to isolation of the worker from his or her network of friends and neighbours. Some workers may no longer be able to participate in hobbies they used to share with friends or may isolate themselves because of feelings of depression.

The New Zealand study involved the analysis of 15 serious occupational injury or illness claims. The study quantified the costs of those accidents and illnesses to the worker, the employer, the government, and private insurers, and came up with a total cost of $1,167,471.84 (about $853,000 CDN); future costs were $3,985,989.00 (about $2,890,000 CDN). Based on these figures, the average total cost per case was over $249,500 CDN.

A Canadian study performed by the CBC and released in 2006 estimated that the annual cost to Canada's economy of "stress-related injuries" is between $16 and 33 *billion*. Clearly, wherever we live, workplace accidents and injuries cost us all.

Government's Perspective

When we consider the impact of occupational injuries and illnesses, the effect on government likely doesn't come to mind first. However, the financial cost of workplace accidents and illnesses to Canadian governments is staggering.

Ontario's occupational health and safety **legislation** is administered by the provincial government—more specifically, by the **Ministry of Labour (MOL)**. The MOL, as you will learn in later chapters of this book, performs workplace inspections and investigates more serious workplace accidents and incidents of illness. The **Ontario Labour Relations Board (OLRB)** is an administrative tribunal that hears a number of different kinds of proceedings under occupational health and safety laws—for example, **appeals** of the orders made by occupational health and safety inspectors.

A serious occupational accident—for example, one that results in death or permanent major disability—typically triggers a full-fledged (and expensive) **prosecution** heard in the Ontario Court of Justice.

Because Ontario has publicly funded health care, the medical costs of injured workers are borne in large part by the government. In some cases, injured workers who are disabled may be entitled to government-funded disability benefits.

legislation
a law passed by a parliament and codified in writing

Ministry of Labour (MOL)
a ministry of the Ontario government that regulates many facets—including occupational health and safety—of labour and employment in the province

Ontario Labour Relations Board (OLRB)
an independent tribunal that mediates, arbitrates, or adjudicates certain disputes relating to labour and employment in Ontario

appeal
the referral of a case to a higher court for a reconsideration of the decision of a lower court

prosecution
a trial of a person in a court of law for a criminal offence

The New Zealand study cited earlier considered the range of costs of occupational injuries to the government; its general findings are quoted below. While the occupational health and safety system in New Zealand is not identical to Ontario's, the "social safety net" in place in New Zealand is similar in scope to Ontario's, and so many of the costs observed there would have counterparts here:

> Costs for the government could be both direct and indirect. There were costs related to infrastructure and services. For instance provision of the justice system, such as courts and the collection of fines. Less obviously, there were costs to the economy through the loss of paid and unpaid work. The government lost taxation revenue. Loss of income through workplace illness and injury can lead to reduced taxable income for the government. ACC's [Accident Compensation Corporation] and OSH's costs were considerable in the study. ACC economic costs were wide-ranging, including anything from the funding of acute health care to rehabilitation to income replacement, and the costs of administration ... There are also social costs in administering legislation. OSH and the ACC are obliged to perform particular functions that are not always well understood or appreciated. ACC staff are required to make difficult cover decisions. OSH investigations, where prosecution is a possibility, can be unpleasant. (Adams et al., 2002, p. 17)

The "social costs" referred to in the New Zealand study included the impact of an occupational accident investigation, from a political perspective, on various parties' perceptions of their government. The study found, not surprisingly, that the government agency that investigated the incidents "was seen as supportive by the participants ... but overreactive by the employer." It also found that the investigations of claims of occupational illness (as opposed to injury) were especially costly and administratively complex because the evidence in these cases (for example, evidence about the precise cause of an illness) was less clear-cut.

As the study shows, investigating workplace accidents and funding the treatment of workplace injuries and illnesses are a significant drain on government resources.

APPROACHES TO WORKPLACE SAFETY

As noted in the Introduction, this is a book about law, and the chapters that follow will focus on the law's response to workplace accidents and injuries. However, it's important to realize that legislating safety is only one of a wide range of strategies for reducing workplace injuries and illness. Disciplines such as engineering and psychology have important roles to play.

Law is to some extent reactionary. Many laws create prescriptions (things that must be done) and prohibitions (things that must not be done) that are backed by offences (what will happen if you do or don't do a particular thing). Even then, negative legal consequences do not always follow non-compliance with safety rules, because rule-breakers are not always caught. This means that the law will never be a completely effective accident and illness prevention tool. By contrast, some engineering innovations actually eliminate the hazards from particular work tasks, while some behaviour-modification strategies create internal motivation to work safely on the part of workers. For these reasons, the best route to reducing occupational injuries and illness is a multidisciplinary approach—one that combines the best contributions of many different disciplines: law, occupational medicine, engineering, industrial hygiene, ergonomics, toxicology, psychology, and human resource management, to name a few.

PROFILE

MACCK Industrial Hygiene Inc.

Industrial hygiene is just one example of a discipline other than law that plays a key role in promoting occupational health and safety. James MacKenzie, CIH of MACCK, answered the following general questions about the discipline and his company's services.

Q: *What is industrial hygiene?*

A: Industrial hygiene involves the anticipation, recognition, evaluation, and control of physical, chemical, and biological agents in the workplace.

Q: *Which professional designations should clients look for in an industrial hygienist?*

A: Designations such as CIH (certified industrial hygienist), and ROH (registered occupational hygienist), are the more recognizable certifications for professional hygienists in the field of industrial hygiene. Technical designations such as OHST (occupational health and safety technologist) and ROHT (registered occupational hygiene technologist) are also common for those in industrial hygiene practice.

Q: *What typically prompts a client to seek MACCK's services?*

A: JHSCs' requests, as well as requests through MOL orders, are the more common avenues by which workplaces initially approach companies like MACCK Industrial Hygiene, Inc. for their health and safety concerns. Many of our clients establish ongoing health and safety programs that include regular industrial hygiene surveys.

Q: *What kinds of services does MACCK provide?*

A: MAACK provides industrial hygiene air monitoring, designated substance assessments [for example, for compliance with the Workplace Hazardous Materials Information System (WHMIS); see chapter 4 of this book], noise surveys, asbestos inspections (including sampling and analysis), indoor air quality surveys, airborne mould spore sampling, ventilation surveys, and laboratory analysis services. The company also arranges for consultations in ergonomics and medical surveillance.

KEY TERMS

claim	legislation
union	Ministry of Labour (MOL)
Occupational Health and Safety Act (OHSA)	Ontario Labour Relations Board (OLRB)
workers' compensation	appeal
benefits	prosecution

REVIEW QUESTIONS

1. How do employees' perspectives with regard to safety usually evolve from the period before an accident occurs to afterward?

2. Why is it difficult for employers to measure the contribution of safe work procedures to the organization's bottom line?

3. When an employee is hurt at work, what consequences might his or her spouse suffer?

4. List four kinds of costs that the government might incur in the wake of a workplace accident or incident of illness.

5. What are the limitations of the law as a tool for promoting workplace safety?

6. List three disciplines, other than law, which offer useful approaches to safety promotion.

DISCUSSION QUESTIONS

1. Why is it useful for human resources professionals to consider employee, union, community, and government perspectives on occupational health and safety?

2. How might a business that is seeking to position itself as a safety leader support community wellness and rehabilitation programs?

3. In this chapter we mentioned the political implications, for government, of safety inspections and prosecutions. Why do you think the MOL (or the Ontario government in general) might care about how its investigation and prosecution activities are perceived by employees? By employers?

CHAPTER 2

The Evolution of Workplace Safety Law

CHAPTER OBJECTIVES

After completing this chapter, you should be able to:

- Explain how workers sought compensation for on-the-job injuries before the passage of worker safety laws.

- Describe some of the earliest workers' compensation initiatives that emerged in Europe before being adopted in Canada.

- Describe the constitutional division of powers (between the federal and provincial governments) as it relates to occupational health and safety.

- List four statutes that relate to the safety of workers in Ontario.

- Explain how environmental laws play a role in supporting worker safety.

INTRODUCTION

While most legal developments in the area of workplace safety began after the Industrial Revolution, workplace injury is as old as work itself.

Consider, for example, the level of risk that workers in long-standing professions such as deep-sea fishing, mining, blacksmithing, and early law enforcement shouldered daily. Workers did not even have to leave home to be at risk—farming has always been one of the most dangerous occupations, and continues to be to this day. The Canadian Agricultural Injury Surveillance Program (CAISP) recorded 1,547 deaths from agricultural injuries between 1990 and 2003, while the 2005 Alberta Farm Injuries Report noted the following for that year:

- 1,353 farm injuries were reported to hospitals in Alberta;

- 41 percent of cases involved injuries related to livestock handling, 22 percent involved machinery, 9 percent involved tools, and 4 percent involved chemicals;

- and 14 percent of all injuries were to children under age 17. (2005 Alberta Farm Injuries Report, 2006)

While getting hurt has always been an occupational risk, workers have often made a distinction between injuries suffered while working for oneself—as a farmer or an independent tradesperson, for example—and injuries suffered while working in the service of another.

In general, employees have less control over the risks of their work than do self-employed people. Where a self-employed person is generally free to avoid work that is not worth the risk—for example, going out fishing in a heavy gale—employees have always been less free to make such judgments without jeopardizing the security of their employment. In difficult economic times in particular, employees have been willing to shoulder considerable risk in order to earn a living.

During the early years of the Industrial Revolution, when large-scale manufacturing techniques were in the earliest stages of development, machines and processes lacked in sophistication. Industrial work was often dirty, noisy, hot, and dangerous. Equipment malfunctions and human error led, on a fairly regular basis, to gruesome and disabling injuries. At the same time, compensation through **litigation** held little promise for victims. In a speech at the 1994 convention of the Association of Workers' Compensation Boards of Canada in Toronto, Dr. Robert D. Elgie, then chair of the Nova Scotia Workers' Compensation Board, summarized the situation as follows:

> Early jurisprudence in the 1800s made it almost impossible for workers to succeed in law suits against their employers in contract or in negligence. A series of very potent defenses which became known as the "unholy trinity," arose—the voluntary acceptance of risk; the fellow worker rule; and the contributory negligence rule—and any or all of these defenses created a virtual bar against any successful action. The late Professor Arthur Larson estimated that since only about 17% of accidents were [found by the courts to be] due to employer "fault," there was little left of employee protection. There were, as well, grave practical difficulties for any worker pursuing litigation—paying for his/her medical treatment; securing and paying legal counsel; and finding the necessary funds to survive while waiting for an offer of settlement [or] a judgment.

In short, the **common law** of **tort**—at the time the only legal route to compensation—was ineffective, and the result was worker and general social unrest. By the end of the 19th century European governments were beginning to consider legislative approaches to the issue.

This chapter will outline the history of workplace safety laws, first in Europe and then in Canada, where the legislation will be placed in a constitutional context. Legislating safety in a coherent, coordinated way has taken over 100 years to accomplish, and yet the legislation continues to evolve; as you saw in chapter 1, the OHSA was extended to cover agricultural workers in June 2006.

COMPENSATION FOR INJURED WORKERS

In Other Jurisdictions

While piecemeal rules relating to safety were included in certain 19th-century **statutes**, a universal approach to worker safety did not emerge in North America until the second half of the 20th century. Interestingly, legislative advances focused first not on prevention but rather on how to compensate workers who had been hurt at work.

---

litigation
legal action, usually with a court component—for example, a lawsuit

common law
the body of law derived from custom and past court decisions rather than statutes

tort
a wrongful act for which damages can be sought in a civil court by the injured party

statute
a written law passed by a parliament

As noted in the Elgie quotation above, the state of the common law at the time made it very difficult for workers to successfully sue their employers for workplace injuries. Under the law, three strong **defences** were available as a means for employers to deny or minimize liability for workplace harm:

1. Contributory negligence: Where an employer could prove that an act or omission of a worker contributed to the worker's injury, the employer was not considered to be at fault in the accident.
2. Fellow-worker rule: Where a co-worker contributed to a worker's injury, this intervening cause relieved the employer of liability for the accident.
3. Voluntary assumption of risk: When a worker accepted work, the law interpreted the worker's acceptance as evidence of the worker's voluntary assumption of the risks inherent in the work, which in turn relieved the employer of liability for those risks.

defence
an argument that the responding party, in a lawsuit, makes to counter the suing party's allegation of fault

The first government to create a system for the compensation of injured workers was Germany, which at the time (1884) was headed by the very conservative Chancellor Otto von Bismarck. The ambitious German approach to worker compensation involved the creation of an insurance-style fund that would provide benefits to injured workers or their survivors. Both employers and employees were required to contribute to the fund—a feature that avoided the pitfalls of other early compensation systems, which tended to bankrupt companies (particularly when several workers were injured in a single accident). The result was a system that, while imposing a greater burden on workers, was superior to the existing British approach that held employers solely responsible for funding compensation. It was not until 1897 that Britain followed suit, introducing its own **no-fault** workers' compensation legislation. Benefits payable to injured workers were low, but employers could no longer rely on common-law defences to avoid assuming some responsibility for worker injuries.

no-fault
(of insurance, etc.) a system under which compensation is available for losses without the need to formally attribute fault for those losses to any particular party

In Canada

Around the same time, Canadian governments had begun to explore the issue of worker compensation. In 1886 Ontario passed the *Workmen's Compensation for Injuries Act*; the statute, however, like the British legislation solely on which it was based, placed the burden of compensating workers on employers.

The federal government was also struggling with the issue. In 1889 a report of the Royal Commission on the Relations of Labour and Capital was released containing recommendations for the creation of a compensation system. The commission, however, was a federal creation, and under Canada's constitution the regulation of labour and employment falls under provincial **jurisdiction**. (The constitutional division of powers is discussed in greater detail under the heading "Legislating Workplace Safety," below.)

jurisdiction
the subject and/or geographical area over which the legal authority of a particular statute or court extends

Unsatisfied with the state of affairs under the *Workmen's Compensation for Injuries Act*, in 1910 Ontario Premier James Pliny Whitney decided that the issue of worker compensation deserved a closer look. He appointed Mr. Justice William Meredith as head of a royal commission to study workers' compensation laws in other jurisdictions. The Meredith report, which was released in 1913, paved the way for the 1914 passage of Ontario's first no-fault worker compensation legislation, the *Workmen's Compensation Act* (WCA).

The passage of the first WCA was a historic step in the recognition of worker rights. In fact, as noted by Dr. Robert Elgie in the same speech quoted earlier in this chapter, workers' compensation was "the first significant social support program to be introduced into this country." While the WCA did not solve the underlying problem of prevalent workplace accidents, it helped ensure that a worker's disabling injury no longer led inevitably to an entire family's plunge into poverty.

※ Success Story

Canadian Physicians Raise Awareness of Occupational Disease

Occupational disease, like **occupational injury**, has been around as long as occupations have. Drawing a distinction between "injuries" and "diseases" acquired at work is somewhat artificial, because the boundaries between the two are not always very clear. For example, acute poisoning—in which the symptoms appear suddenly in a near-immediate reaction to a spill—is classified by most workplaces as an occupational injury. However, chronic diseases such as silicosis (which take years or even decades to appear) are classified as diseases, even though, like acute poisoning, they represent the direct effects of a workplace exposure.

A key difference between what we tend to think of as "injuries" and "diseases" is that, while it's easy to point to the causes of a workplace injury (because the harm happens instantly and there are usually witnesses), diagnosing and establishing the precise causes of a disease is much more complicated. Many diseases related to occupational exposures can also be caused by exposures in the home or in other locations, and many symptoms of recognized occupational diseases are shared by other diseases that are not related to workplace risks.

These evidentiary difficulties have meant that efforts to recognize, provide compensation for, and prevent occupational diseases have lagged behind injury-prevention innovations.

However, when a number of people who work in the same industry find themselves afflicted with the same pattern of symptoms, the situation becomes difficult to ignore. An example of this occurred in the middle of the 19th century, when scores of coal workers were diagnosed with pneumoconiosis, or "black lung." In 1876 prominent Canadian physician Sir William Osler wrote a paper on pneumoconiosis. Other exposures that received attention near the turn of the 20th century in Canada were to asbestos, western red cedar, and silica.

A leading researcher in the area of silica—and silicosis, an occupational disease caused by silica exposure—was Dr. J. Grant Cunningham. In 1920 Dr. Cunningham established the Industrial Health Branch of the federal government. As director of that office, Cunningham spent 40 years working to raise awareness of the issue of occupational disease, both in Canada and internationally.

Today, every province in Canada provides compensation for victims of occupational disease in its legislation. At the federal level of government, the Canadian Centre for Occupational Health and Safety (CCOHS) continues the work begun by the (now defunct) Industrial Health Branch by providing information on occupational diseases for industry and workers.

All occupational diseases are preventable: if a worker is not exposed, he or she will not get sick. This simple fact underscores the importance of doing as much as possible to avoid exposing workers to dangerous substances in the workplace. The work of Sir William Osler, Dr. J. Grant Cunningham, and others like them paved the way for what will be, for industry, an ongoing quest to make working in Canadian industry safe.

---

occupational disease
in broad terms, any disease that ensues after a workplace exposure and that has been demonstrated to be associated with that type of exposure

occupational injury
any injury that occurs while working, or while at one's workplace

LEGISLATING WORKPLACE SAFETY

Introduction

As of January 1, 1915—the date the WCA came into force—Ontario workers could count on a certain percentage of income replacement in the event of a serious work injury. Workplace catastrophes, however, continued to claim the lives of Ontarians, especially in dangerous industries such as mining and construction.

In March 1960 five immigrant workers were killed when the water main tunnel they were building in Toronto's Hogg's Hollow neighbourhood collapsed. The incident sparked a public demand for a review of worker safety laws, and the province responded by creating a royal commission to study the issue. The Hogg's Hollow accident and the commission report reawakened discussions about the question of jurisdiction over safety laws in Canada.

HARD LESSONS
The Westray Mine Disaster

The Westray mine was a coal mine near the village of Plymouth, Nova Scotia. The mine was established on the Foord seam—a rich, underground coal deposit.

Coal seams typically harbour not only coal, but also various gases, including methane gas. The Foord seam contains exceptionally high levels of this gas. When trapped—for example, in an underground (closed instead of open-pit) mine—methane gas has a high potential for explosion.

Before the Westray disaster, the Foord seam and the mines on it (of which there have been at least eight) had already developed a reputation for danger. Between 1866 and 1972 there were about 600 mining-related deaths in the mines of Pictou county, and another mine in the region—the Allan mine—had experienced eight methane explosions in its operating history.

Construction of the Westray mine was supported by considerable government funding from both the federal and provincial levels, and before the mine was built its founders had already arranged a significant supply contract with Nova Scotia's most important power company. The mine was promoted as a facility with "state-of-the-art" technology, both in terms of production and safety; however, serious ground control problems plagued it from the start of operations in the fall of 1991.

At 5:18 a.m. on the morning of May 9, 1992 the mine was destroyed in a massive explosion. The top of the mine entrance was blown off, steel supports throughout the mine collapsed, and windows shattered in houses in nearby villages. All 26 workers who were underground at the time were killed in the blast or in the resulting collapse.

The inquiry into the incident revealed that at the time of the accident a miner was using a "continuous miner" (a machine that uses a rotating drum and tungsten carbide teeth to scrape coal off of surfaces) when the machine struck some pyrite, a hard mineral. This caused a shower of sparks, which ignited methane gas present in the area and created a rolling methane flame that travelled through the mine corridor. The flame gained momentum until it caused a methane explosion, followed by a high-pressure shockwave that triggered a massive coal dust explosion. Most of the miners were killed by the shockwave; the remainder died of carbon monoxide poisoning when the fire consumed all of the available oxygen.

In the aftermath of the disaster, media reports suggested that miners working in the mine suspected that the work was unusually dangerous but that they were sometimes pressured to work in unacceptable conditions. The scarcity of stable work in that part of Nova Scotia at the time also likely exerted considerable economic pressure on miners to accept jobs at Westray and to remain in those jobs, despite the knowledge that the work was very risky.

The Westray disaster was the subject of legal proceedings in both the civil and the criminal courts.

On November 7, 2003 the federal House of Commons gave royal assent to (made law) Bill C-45, *An Act to Amend the Criminal Code (Criminal Liability of Corporations)*. The Bill, which was inspired by the lessons learned from the Westray disaster, made changes to the **Criminal Code** to make it easier for corporations to be criminally charged for disregarding safety warnings.

Before these amendments, certain corporate officers at a very senior level could be charged with criminal **negligence** related to workplace accidents. However, the charges could only be laid when it was proven that a person who was a "directing mind" of the corporation engaged personally in conduct that amounted to criminal negligence. From a practical standpoint, very senior staff in large corporations like Westray are not usually sufficiently involved with day-to-day business or safety operations to be in a position to be personally negligent.

Bill C-45 changed the law so that, today:

- a corporation can be held liable for the criminally negligent acts of lower-level staff who exercise "delegated operational authority"; and

- a corporation can be charged criminally in its own right (that is, as a corporation) instead of only on the basis of the criminal negligence of its directing minds (see *Criminal Code* sections 22.1 and 22.2).

The legislation also added a new provision to the *Criminal Code* that makes it clear that anyone who directs the work of another has a legal duty to take reasonable steps to protect that person from bodily harm:

> 217.1 Everyone who undertakes, or has the authority, to direct how another person does work or performs a task is under a legal duty to take reasonable steps to prevent bodily harm to that person, or any other person, arising from that work or task.

Criminal Code
the statute that describes the legislative component of Canada's criminal law

negligence
the failure of a person to respect or carry out a duty of care owed to another

The Constitutional Context of Safety Legislation

Canada's government, as created by the *Constitution Act, 1867*, is a federation. Canada's federal system creates two main levels of law-making authority: the federal government at the first level, and the provincial and territorial governments at the second level. When it was drafted, our constitution divided the areas of regulatory authority that were deemed important at the time into two groups. The areas (or subject matters) in the first group were assigned to the federal government under section 91, and those in the second group were assigned to the provincial governments under section 92.

Section 92(16) of the constitution assigns to the provinces jurisdiction over "Generally all Matters of a local or private Nature in the Province." This provision is quite vague, but had been interpreted, by the federal government, to include the regulation of labour and employment within the provinces. This interpretation was the reason why the federal government, in 1889, considered workers' compensation or workplace safety laws to be outside its jurisdiction.

The constitution, however, did grant the federal government jurisdiction to regulate a wide range of industries considered to be of national importance, including:

- postal service (section 91(5));
- the militia, military, naval service and national defence (section 91(7));
- navigation and shipping (section 91(10));
- fisheries (section 91(12));
- banking (section 91(15));
- patent and copyright management (sections 91(22)–(23)); and
- federal penitentiaries (section 91(28)), among others.

The constitution also gives the federal government jurisdiction over "the fixing and providing for the Salaries and Allowances of Civil and other Officers of the Government of Canada" (section 91(8)), which has been interpreted to mean that the federal government is entitled to regulate federal employees. In the first half of the 20th century, the federal government accomplished this via a patchwork of statutes. In 1967 these statutes were consolidated and became the **Canada Labour Code**.

In 1968 the federal government clarified the constitutional context for workplace safety legislation by defining its own jurisdiction over occupational health and safety for federal workers, and granting jurisdiction to the provinces to pass safety laws governing other employees.

The first province to pass a general (multi-industry) workplace safety law was Saskatchewan. Its 1972 legislation is considered to be the first of its kind in North America. The other Canadian provinces soon followed suit, with Ontario introducing the OHSA in 1978.

Canada Labour Code
a federal statute that deals with labour and employment issues, including occupational health and safety, that affect employees of the federal government or of certain federal works

ENVIRONMENTAL PROTECTION AND WORKPLACE SAFETY

Introduction

While not directly worker-focused, laws relating to the protection of the environment have had a very important impact on work and workers.

Human beings and their activities are part of the world's ecology. Like every other species, we eat, seek shelter and warmth, produce waste, reproduce, and die. Many industries, such as farming, logging, and fishing, can be thought of as extensions of these basic functions. However, human beings sometimes create a disproportionate and harmful impact on the natural world—an impact that is felt most often when we use industrial innovations like machines, processes, and chemicals to increase the rate at which we harvest the world's resources and convert them into forms that suit us.

Where harm to the natural environment begins to affect our neighbours' enjoyment of their land or when we witness negative effects on plants, animals, and fellow

FACTS AND FIGURES

OHS in Ontario—Timeline

1884   Germany establishes the first insurance-model workers' compensation fund.

1886   Ontario passes the *Workmen's Compensation for Injuries Act*, a statute based on a similar British law. This law made compensation of injured workers compulsory, but only after a legal finding of fault. Under this system, employers were solely responsible for payment of claims.

1889   The federal Royal Commission on the Relations of Labour and Capital recommends legislative action to improve working conditions in dangerous industries; however, the federal government declines to act on the recommendations, believing that doing so would infringe on provincial legislative jurisdiction under the constitution.

1897   Britain introduces the *Workmen's Compensation Act*, its first no-fault workers' compensation scheme. Benefits payable under the Act are low, but compulsory.

1910   Ontario Premier James Pliny Whitney appoints Mr. Justice William Meredith as head of a royal commission to study workers' compensation laws in other jurisdictions.

1913   The Meredith report is released and recommends the introduction of a scheme under which employees give up their rights to sue employers in the wake of an accident in exchange for guaranteed protection for workers against post-accident income loss regardless of fault.

1914   Ontario passes the WCA, the province's first workplace accident/illness insurance scheme. It comes into force on January 1, 1915.

1920   The Ontario Board of Health establishes an industrial hygiene division designed to monitor and detect occupational diseases and promote their prevention.

1960   After five workers are killed while working on a tunnel in Toronto, the Ontario government forms a commission charged with the task of recommending updates to worker safety laws.

1968   The *Canada Labour Code* grants the provinces legislative authority to pass worker safety laws in many industries (although many provinces had already begun to do so within their workers' compensation laws).

1972   Saskatchewan passes an occupational health and safety statute, considered to be the first such statute in North America.

1974   Ontario appoints a royal commission headed by James Ham to make recommendations for worker safety laws.

1976   The Ham commission report is published and recommends an approach to safety that includes the creation of an internal responsibility system and the right of workers to know about safety hazards at work, to participate in safety initiatives, and to refuse unsafe work without penalty.

1978   Ontario introduces the OHSA, which comes into force in 1979.

1986   A *Canada Labour Code* amendment introduces joint management–employee health and safety committees.

1988   A collaborative effort by the federal and provincial governments results in the creation of the WHMIS.

humans, we tend to look for ways to avoid those consequences. Passing laws to limit harmful behaviour or to prescribe protective behaviour has been one way in which the world's governments have responded to environmental harm. These laws have in turn required industries to change their practices, which often affects the lives of workers.

For example, the tannery (leather processing and dyeing) industry was once infamous for its use of highly toxic chemicals. Tannery workers were forced to work in close proximity to powerful fumes and to touch materials that posed serious chemical and biological hazards. Today, the introduction of automation of many tannery tasks and the development of less-toxic chemicals and solvents has led to the following improvements:

- fewer toxic emissions into the air, soil, and water in communities near tanneries;
- leather products that are less risky for consumers; and
- much safer working conditions for tannery workers.

Environmental laws that regulate tannery effluents are just one example of the influence of environmental law on workplace safety.

Overview of Environmental Law in Canada

Before the advent of environmental protection legislation, people who suffered property damage or other harm (for example, physical harm or business losses) due to another party's polluting activities were generally forced to seek a remedy in the civil courts, under the common law.

One branch of the common law is the law of torts, or wrongs. Under the law of torts, a person or company can bring a **lawsuit** against another person or company, alleging, for example, negligence or nuisance—two subdivisions within the law of torts.

The law of nuisance was used, historically, to deal with a wide range of disputes that most people now consider to be "environmental law." For example, people whose properties were affected by another landowner's emissions (of smoke, waste water, bad odours, and so on) were, before the advent of environmental law, restricted in their search for a remedy to a tort lawsuit in the civil courts. An example of an early suit of this nature might be where one business (say, a holiday resort) sues when the business beside it begins undertaking an activity (say, a smelly pulp mill) that affects its ability to attract customers.

As a tool for resolving environmental problems, tort law had some serious limitations. First of all, few plaintiffs had the financial means to sue, and many would instead opt to give up and move away from a polluter. Second, the law of tort requires strict proof of the cause of harm that is alleged. Some deleterious effects on a property—for instance, soil contamination—can have multiple causes, especially in an industrial area. In some cases, contamination of soil is historical, which means that by the time the contamination is discovered, the polluter may have long since moved away. For the party affected, it could be extremely difficult to prove exactly which neighbour or neighbours, past or present, caused the harm.

By the 1960s Canadian parliaments had begun to address these kinds of problems in legislation. In Canada, constitutional jurisdiction over the environment is

lawsuit
a formal request, by one party, for the court's assistance in obtaining, through its processes, compensation or relief as against another party

shared. The provinces have jurisdiction over the management of the natural resources within their boundaries and over many kinds of industry and commerce, while the federal government has jurisdiction over some interprovincial industries (for example, railways and fisheries) and also has the power to make laws in support of "peace, order and good government" in the nation.

At the federal level, the most general piece of environmental legislation is the *Canadian Environmental Protection Act* (CEPA). Under this Act, there are many **regulations** that deal with more specific topics. The federal government also regulates environmental issues in the Arctic and in the oceans and seas that border on Canadian lands.

At the provincial level, there are literally hundreds of different environmental laws. Some, like the Ontario *Environmental Protection Act*, deal with environmental regulation from a general perspective; others, like the *Safe Drinking Water Act*, deal with individual environmental issues. Other pieces of legislation include environmental protection provisions under a more general scheme of regulation; the *Drug and Pharmacies Regulation Act*, for example, prescribes rules for the disposal of pharmaceutical waste by pharmacies, laboratories, and manufacturers.

Though they are not workplace safety laws per se, many of these laws were drafted with the protection of workers in mind. While a polluting substance may be harmful to plants, animals, and people when it escapes into the environment, the people who work with the substance every day sustain a much higher level of exposure. For this reason, laws that protect the environment protect workers, too. For employers, full compliance with workplace safety laws can rarely be achieved unless environmental law compliance is achieved as well.

regulation
a legal instrument that is subservient to a statute and created to provide guidance for the administration of the statute

☀ Success Story

Environmentalists Urge Alternatives to Tetrachloroethylene Use in Dry-Cleaning Operations

Tetrachloroethylene, also known as perchloroethylene or PERC, has long been used in dry-cleaning operations across Canada. However, PERC has significant toxic effects, not only on the environment in general but on the health of dry-cleaning consumers and employees of dry cleaners.

According to the CCOHS tetrachloroethylene is a severe skin irritant, an eye irritant, and a serious airway irritant. Short-term exposures (a few hours at moderate levels) can cause drowsiness, dizziness, headache, nausea, loss of coordination, confusion, and unconsciousness. Persistent exposures (for at least four hours a day for several days) can cause behavioural changes such as slow reaction to visual stimuli and loss of coordination. Exposures to very high vapour concentrations have caused deaths. Over the long term, it is suspected that tetrachloroethylene exposure causes harmful effects on the central nervous system, the heart, the kidneys, and the liver; it may also cause esophageal cancer.

Increased awareness of the harmful environmental effects of tetrachloroethylene in the 1990s led the federal government to undertake a stakeholder consultation process focusing on how to manage the use of the chemical in the dry-cleaning industry. A 1996 report on the results of this process led the government to create the Tetrachloroethylene (Use in Dry Cleaning and Reporting Requirements) Regulations (under the CEPA).

These regulations:

- banned the use of tetrachloroethylene in spotting agents (substances applied directly to stains by dry-cleaning workers);
- required that tetrachloroethylene, including waste water contaminated with the chemical, be stored by dry cleaners in closed containers;
- required the replacement of older dry-cleaning machines with newer machines that obtain the same results using far less tetrachloroethylene, and that incorporate special containment features;
- imposed new waste management and reporting obligations on the dry-cleaning industry and suppliers; and
- set a target for tetrachloroethylene use in the dry-cleaning industry that is 71 percent lower than the estimated use in 1994.

The regulations, which should significantly reduce tetrachloroethylene exposure for both the general public and dry-cleaning workers, came into force in 2003.

KEY TERMS

litigation	occupational disease
common law	occupational injury
tort	*Criminal Code*
statute	negligence
defence	*Canada Labour Code*
no-fault	lawsuit
jurisdiction	regulation

REVIEW QUESTIONS

1. How did workers obtain compensation for workplace injuries before the advent of workers' compensation laws?

2. How does a "no-fault" compensation system differ from a fault-based system? Why do you think Ontario has adopted a worker compensation system based on no-fault principles?

3. Before 1978, where would you have had to look for Ontario workplace safety laws?

4. Are federal government workers working in Ontario governed by the OHSA? If not, which law(s) govern the safety of these workers?

5. What is an occupational disease?

6. How do environmental laws affect workplace safety for employees?

DISCUSSION QUESTIONS

1. Ontario's worker compensation legislation is based on an insurance-style model. Both employees and employers pay into the insurance fund. Why do you think the system's designers wanted employees to contribute? Do you agree with this aspect of the system?

2. Why is it important for occupational health and safety to be dealt with under a general "umbrella" statute like the OHSA instead of via provisions in the legislation that governs individual industries?

CHAPTER 3

Overview of the OHSA

CHAPTER OBJECTIVES

After completing this chapter, you should be able to:

- Understand the goals and basic philosophy of the OHSA.
- Describe the general structure of the OHSA.
- Explain to which kinds of workplaces the OHSA applies.
- Know which government ministry enforces the OHSA and describe the powers of Ministry inspectors and Directors.
- Explain the procedure for work stoppages and refusals.
- Understand the philosophy and purpose behind the Internal Responsibility System (IRS) created by the OHSA.
- Describe the roles of employers/constructors, supervisors, unions, and workers under the IRS.
- Understand the role and duties of Health and Safety Representatives and Joint Health and Safety Committees (JHSCs) within the IRS.

INTRODUCTION

As you learned in chapter 2, the Ontario legislature's passage of the 1978 OHSA represented the province's first attempt to address safety at work as a general—as opposed to an industry-specific—issue. While some statutes are passed as a means of putting a collection of related common-law rules into print, others have the more ambitious goal of creating a system—or framework—for human interaction.

Examples of this kind of "framework legislation" that may be familiar to you include the *Family Law Act*, which prescribes a plan for addressing various issues (such as property division, custody, and support) that arise when a couple separates, and the *Compulsory Automobile Insurance Act*, which creates a scheme for ensuring that people who suffer injuries or property damage in a car accident are compensated while minimizing the need for these people to resort to the already burdened legal system.

The OHSA is another excellent example of scheme-creating legislation. As you will learn in this chapter, it adopts the philosophy that workplace safety is a shared responsibility. In giving rights to workers (for example, the right to refuse unsafe work), the OHSA explicitly recognizes the value of worker participation and employer responsiveness to the process of creating a safe workplace. As a tool for reinforcing its philosophy, the OHSA defines the legal duties and responsibilities of the key workplace parties—workers, management, unions, and government regulators.

The OHSA creates both an internal responsibility system, made up of the various roles and responsibilities of the workplace parties, and an external responsibility system, in the form of supervisory and enforcement powers assigned to the MOL.

This book is written primarily from a management perspective. For that reason, its primary focus will be on identifying and explaining employer compliance requirements. However, it is important for managers to understand the occupational health and safety roles of other parties—employees, unions, and government—because such an understanding is an essential foundation for building good working relationships with each of these groups. This chapter will provide an overview of the structure of the OHSA, followed by a description of the **Internal Responsibility System (IRS)** and the roles of the parties within it.

Internal Responsibility System (IRS)
a scheme, created under the OHSA, that assigns health and safety responsibilities to each workplace party

STRUCTURE OF THE OHSA

Table of Contents

This section provides a summary of the structure of the OHSA in order to give you an idea of what the legislation does (and does not) do. A basic table of contents for the OHSA is reproduced in figure 3.1.

Definitions

Many statutes begin with a Definitions provision. In the OHSA, this provision (section 1) appears before the parts that subdivide the rest of the statute. There are 35 defined terms in the OHSA.

The definitions that a statute ascribes to the terms that appear in it are extremely important. Legislators define terms for a variety of reasons, including:

Figure 3.1 Occupational Health and Safety Act, RSO 1990, c. O.1

Contents:	(contains only section 1, Definitions)
Part I:	Application
Part II:	Administration
Part III:	Duties of Employers and Other Persons
Part III.I:	Codes of Practice
Part IV:	Toxic Substances
Part V:	Right to Refuse or Stop Work Where Health or Safety in Danger
Part VI:	Reprisals by Employer Prohibited
Part VII:	Notices
Part VIII:	Enforcement
Part IX:	Offences and Penalties
Part X:	Regulations

- to establish a "short form" for referring to certain people, places, or things. For example, section 1 defines "Board" as the Labour Relations Board, and "Minister" as the Minister of Labour. This is important, because there are many boards and ministers in Ontario who do *not* have roles to play under the OHSA.

- to clarify the precise meaning of technical or legal terms (sometimes called "terms of art"). For example, the OHSA defines "occupational illness" as "a condition that results from exposure in a workplace to a physical, chemical or biological agent to the extent that the normal physiological mechanisms are affected and the health of the worker is impaired thereby and includes an occupational disease for which a worker is entitled to benefits under the *Workplace Safety and Insurance Act, 1997*."

- to limit the application of certain provisions by excluding things from a definition. For example, the OHSA limits the definition of "construction" by saying that construction "includes erection, alteration, repair, dismantling, demolition, structural maintenance, painting, land clearing, earth moving, grading, excavating, trenching, digging, boring, drilling, blasting, or concreting, the installation of any machinery or plant, and any work or undertaking in connection with a project *but does not include any work or undertaking underground in a mine*" [author's emphasis].

If you will be working closely with the OHSA, you should read the Definitions section thoroughly. This will give you an understanding of whether or not the defined terms in the legislation apply to your workplace.

Application

Part I of the OHSA provides that—in addition to the private sector—the OHSA applies to the Crown and its employees; this means Ontario government workers. As you learned in chapter 2 workplace safety for federal government workers is governed by the *Canada Labour Code*.

Part I also makes it clear that if a provision of any other (Ontario) statute conflicts with a provision of the OHSA, the OHSA provision is the one that applies.

Finally, this part describes the exceptions to the application of the OHSA. These include:

- work done on the premises of a private residence by the owner, occupant, or a servant of the owner or occupant of a private residence;

- the work of teachers as defined by the *Education Act*;

- the work of academic staff at universities;

- work done on a farm; and

- with several exceptions, work done by a self-employed person.

This part is somewhat misleading, however, because if you read the regulations made under the OSHA, regulation 857 (RRO 1990) extends the application of the legislation to teachers; as we noted in chapter 1, a new regulation (O. Reg. 414/05) extending the application of the OHSA to agricultural workers came into force in

June 2006. In reality, the OHSA applies to almost *any* workplace in Ontario. In its general guide to the legislation, the MOL has the following to say about the "workplaces" to which the legislation applies:

> A workplace could be a building, a mine, a construction site, an open field, a road, a forest or even a beach. The test is: Is the worker being directed and paid to be there, or to be near there? If the answer is "yes," then it is a workplace. (Ontario Ministry of Labour, 2002a, p. 6)

Administration

Part II of the OHSA deals with administration of occupational health and safety.

ADMINISTRATION WITHIN THE MINISTRY

Sections 5 through 7 of this part give permission to the MOL to designate health and safety inspectors as well as one or more Directors for the purpose of managing occupational health and safety. In some instances, these Directors and inspectors are entitled to exercise powers conferred under the *Canada Labour Code* (see OHSA section 22.1). The government provided for this in an effort to harmonize the administration of some occupational health and safety services.

Section 21 allows the Ministry to appoint advisers or advisory committees to "inquire into and report to the Minister on any matter that the Minister considers advisable." For example, in 2005 the Ergonomics Sub-Committee of the Manufacturing Panel, Health and Safety Action Group presented its recommendations to the Minister of Labour in the form of a report entitled *Recommendations on Strategies to Reduce Work-Related Pains and Strains in Ontario*.

Section 22 makes it clear that the payments that certain employers (Schedule 1 and 2 employers, about which you will learn more in chapter 5) make under the *Workplace Safety and Insurance Act* (WSIA) are allocated in part toward defraying the cost of the administration of the OHSA.

HEALTH AND SAFETY REPRESENTATIVES AND JOINT HEALTH AND SAFETY COMMITTEES

Section 8 provides that every employer or constructor (defined in the Definitions section) with a number of employees that "regularly exceeds five" must appoint a **Health and Safety Representative**. If an employer or constructor does not do this, the Ministry can order that it be done.

The obligation to appoint a single Health and Safety Representative applies only in workplaces where the more onerous obligation to appoint a **Joint Health and Safety Committee (JHSC)** does *not* apply. This usually means workplaces with fewer than 20 employees and construction projects lasting less than three months to which no designated substance regulations apply.

Section 9 imposes an obligation to appoint a JHSC where:

- the employer or constructor has 20 or more employees;
- a designated substance regulation applies to an employer or constructor;

Health and Safety Representative in small workplaces, an individual who is assigned enhanced health and safety rights and responsibilities to be exercised on behalf of workers

Joint Health and Safety Committee (JHSC) in medium and large workplaces, a team of individuals from management and workers, who are assigned enhanced health and safety rights and responsibilities

- a construction project is expected to last longer than three months; and/or

- the employer or constructor is subject to an order, issued by a Director, in response to a concern about a dangerous substance in use at the workplace (section 33 order).

The role and responsibilities of Health and Safety Representatives and JHSCs will be discussed in more detail below.

Section 10 requires constructors of projects lasting at least three months and regularly employing at least 50 employees to establish a Worker Trades Committee. This committee, made up of members from each of the trades represented in the project, reports to the JHSC on worker health and safety concerns.

Duties of Employers and Other Persons

Part III of the OHSA imposes specific occupational health and safety duties on constructors, licensees, employers, supervisors, workers, owners, suppliers, architects, engineers, and corporate directors. These duties will be discussed in detail under the heading "The Internal Responsibility System" later in this chapter.

Codes of Practice

Part III.I, added in 2001 with the passage of Bill 57, gives explicit recognition to the use of Codes of Practice within a workplace as a means of complying with OHSA requirements.

A Code of Practice is a set of guidelines, typically developed by experts in a particular area of practice (for example, the National Fire Protection Association (NFPA)), to direct behaviour—in this case, safety behaviour. Part III.I allows the Minister to approve (or withdraw approval for) a particular Code of Practice by publishing notice of the approval in *The Ontario Gazette* (an official record of government business where new regulations are first officially published).

Once the Minister has approved a Code of Practice, employers who are required to comply with regulations under the OHSA (for example, the Industrial Establishments regulation (RRO 1990, reg. 851)) will be deemed to be in compliance with a provision in the regulations as long as they comply with a Code of Practice that has been approved by the Minister, if that Code covers the same subject matter as the regulation. For example, section 22 of the Industrial Establishments regulation deals with the handling of flammable liquids. NFPA-30 is a Code of Practice that provides guidelines for handling flammable and combustible liquids, such as gasoline. If the Ministry approved NFPA-30, an employer would be deemed to be in compliance with section 22 of the Industrial Establishments regulation by adhering to NFPA-30.

This new part is not uncontroversial. In a May 2001 press release the Ontario Public Service Employees Union (OPSEU) had the following to say about Bill 57's then-proposed change to the OHSA:

> A code of practice is a set of health and safety *guidelines* only. The amendment says that compliance with a code of practice can be seen as compliance with a regulation. However, failure to comply with a code of practice would not be, in itself, a breach of the regulation!

This is a clear weakening of enforcement. Last year, under intense political pressure after the Walkerton tragedy, the Ontario Cabinet toughened the Ontario Drinking Water Guidelines into regulations. Bill 57 goes the opposite direction. It turns strong regulations into weak guidelines and wipes out employer accountability for worker health and safety.

Toxic Substances

Part IV of the OHSA contains provisions that regulate the use of toxic substances in the workplace.

Where a Director (from the Ministry) believes that a substance or combination of substances or agents being used in a workplace is likely to endanger the health of workers, the Director can issue an order to the employer. The order can either prohibit the use of the substance entirely or impose restrictions and conditions on its use.

This part prohibits the manufacture of any new toxic substances at a workplace unless notice of the employer's intention to do so has been given to the Ministry.

Section 35, which deals with designated substances, also appears in this part of the OHSA. Designated substances are managed under the **Workplace Hazardous Materials Information System (WHMIS)**; the WHMIS is discussed in detail in chapter 4 of this book.

Workplace Hazardous Materials Information System (WHMIS)
a coordinated system, supported by legislation at both the federal and provincial levels, for the safe management of hazardous materials in workplaces

Right to Refuse or Stop Work Where Health or Safety in Danger

Part V of the OHSA contains provisions that most people consider to be at the heart of the legislation. The first of these is the right of workers to refuse unsafe work.

The right is expressed at section 43(3) in the following terms:

A worker may refuse to work or do particular work where he or she has reason to believe that,
   (a) any equipment, machine, device or thing the worker is to use or operate is likely to endanger himself, herself or another worker;
   (b) the physical condition of the workplace or the part thereof in which he or she works or is to work is likely to endanger himself or herself; or
   (c) any equipment, machine, device or thing he or she is to use or operate or the physical condition of the workplace or the part thereof in which he or she works or is to work is in contravention of this Act or the regulations and such contravention is likely to endanger himself, herself or another worker.

The right to refuse work, however, is not universal. Certain kinds of workers whose work is considered an essential service do not have the right to refuse work. Examples include work

- where the alleged danger is inherent in or is a normal condition of the work (for example, a hydro worker whose job description requires her to climb hydro poles to fix wiring problems cannot refuse to do this on the basis that climbing hydro poles is dangerous); or
- where the worker's refusal to work would "directly endanger the life, health or safety of another person" (for example, where a correctional officer's refusal to intervene to calm a riot could result in bodily harm to inmates or colleagues).

The workers to whom these limits apply are:

- police officers;
- firefighters;
- most correctional workers;
- hospital, including mental health facility, workers;
- workers in nursing homes and homes for the aged;
- workers in residential facilities or group homes for people with behavioural problems or physical or cognitive disabilities;
- ambulance workers and certain first-aiders;
- some laboratory workers; and
- certain support workers (food service, laundry, power workers, technical workers) in the facilities included in this list.

Workers employed in a facility of the kind mentioned in the above list or in a facility that performs similar services should find out whether any employees in their facility fall within the class of employees who are limited in their right to refuse work. If the employer does not know the answer to this question, a worker should contact the MOL for an opinion. Understanding who can and who cannot refuse work, and under what circumstances, is an important part of a workplace's health and safety policy.

Part V explains what should happen when a person refuses to work. The facts and figures feature below summarizes the appropriate steps.

FACTS AND FIGURES

What Happens When Someone Refuses to Work?

1. The worker who is refusing to work must report his refusal to his supervisor (or to the employer directly).
2. The supervisor or employer's representative (manager, etc.—a particular person may be designated to handle work refusals) should arrange for the attendance of:
   - the Health and Safety Representative;
   - a worker member from the JHSC; or
   - the member of the Worker Trades Committee who represents the worker's trade.
3. The employer and the health and safety person should jointly conduct an investigation of the circumstances of the work refusal "forthwith"—immediately—and in the worker's presence.
4. The employer should remedy the dangerous circumstance so that it no longer poses a danger.
5. If the worker is satisfied, he should go back to work. If the worker disagrees that the work has been made safe, he should inform the employer of this.
6. If the worker continues to refuse to work, either the worker or the employer should inform an inspector from the Ministry.
7. The inspector should attend at the workplace to investigate the circumstances of the work refusal.
8. While waiting for the inspection and the inspector's decision, no other worker should be assigned to do the work that the first worker refused to do.

As you will learn later, an employer can never take any kind of punitive action (known as "**reprisals**") against a worker because the worker has exercised one of his rights under the OHSA—including the right to refuse work.

reprisal
in the context of occupational health and safety, any reprimand, penalty, or negative consequence imposed on an employee by an employer in response to the employee's excercise of OHS rights

Workers directly affected by a potentially unsafe situation are not the only ones who can initiate a safety-related work stoppage. Certified members of a JHSC can also request that work be stopped. Where a single member of a JHSC initiates a work stoppage and investigation, the action is called a "unilateral work stoppage" (see section 47); where members from both parties (worker and employer members) of a JHSC jointly request a work stoppage and investigation, this is called a "bilateral work stoppage." The procedure for bilateral work stoppages is described at section 45.

In cases where a certified member suspects that the bilateral work stoppage procedure will not effectively protect the workers—for example, where the employer is continuing to pressure workers to work under unsafe conditions—the certified member or a Ministry inspector who has become involved can request that a Director issue a "declaration or recommendation" requiring compliance. In the most serious cases, an inspector may be appointed by the Ministry to temporarily oversee health and safety practices at the workplace.

Reprisals by Employer Prohibited

Part VI of the OHSA contains a single provision—section 50, which reads in part:

(1) No employer or person acting on behalf of an employer shall,
   (a) dismiss or threaten to dismiss a worker;
   (b) discipline or suspend or threaten to discipline or suspend a worker;
   (c) impose any penalty upon a worker; or
   (d) intimidate or coerce a worker,

because the worker has acted in compliance with this Act or the regulations or an order made thereunder, has sought the enforcement of this Act or the regulations or has given evidence in a proceeding in respect of the enforcement of this Act or the regulations or in an inquest under the *Coroners Act*.

This provision makes it clear that an employer cannot retaliate in any way against a worker for the worker's attempts to exercise an OHSA right.

Worker complaints about employer reprisals are one of the most common kinds of hearings that come before the OLRB; an example of such a case appears in chapter 4. Employers who are found to have made reprisals against a worker may be subject to a range of penalties, including having to reinstate a fired worker, paying damages to make up for not having given proper notice, and paying fines to the Ministry.

Notices

Sections 51 to 53 form part VII of the OHSA. They describe the procedures employers and constructors must follow in the aftermath of a workplace accident or the diagnosis of an occupational illness. These procedures will be discussed in some detail in chapter 9, which deals with post-incident compliance.

Enforcement

As you learned in the previous chapter, occupational health and safety laws are enforced primarily by provincial governments, and in Ontario, by health and safety inspectors and Directors at the MOL. Occupational health and safety matters that require adjudication (that is, that require a formal decision—for example, when an employer seeks to appeal a Director's order) are dealt with by the OLRB.

Part VIII of the OHSA describes how the Act is enforced by MOL staff. It confers significant powers on inspectors and Directors, for example, by allowing them to:

- enter any workplace without notice and without a **warrant**;

- examine machinery, equipment, and documents, and request access to documents;

- question employees;

- take certain documents out of the workplace (receipts are given);

- conduct tests on equipment or substances, and take substances out of the workplace to run tests;

- make orders requiring employers to remedy a safety problem (Directors only, not inspectors, do this); and

- post a notice of any order made against the employer.

warrant
an order, issued by a court or court officer (such as a justice of the peace), permitting an activity (such as an entry and search) that might otherwise be refused by the party against which it is exercised

Ministry orders can require employers or constructors to do many things. The latter may be obligated to:

- perform or arrange (and pay) for the performance of equipment or laboratory tests;

- produce documents for MOL inspection;

- obtain, at their own expense, engineering reports about safety issues (for example, the load limit of a floor or the ground stability of a mine);

- obtain, again at their own expense, a report on the composition, properties, and/or toxicology of a substance;

- comply with a specific safety regulation; or

- discontinue the use of equipment or substances pending MOL investigation or until a safety hazard is eliminated.

Part VIII of the OHSA provides a procedure (at section 61) under which an employer or constructor can appeal an order that the employer believes is not warranted. This process is explained in chapter 10 of this book.

When an inspector comes to a workplace, the employer must allow the Health and Safety Representative or a worker-side member of the JHSC the opportunity to accompany the inspector on her inspection. In lieu of attending at the workplace herself, an inspector can, unless a provision requires otherwise, order the workplace's Health and Safety Representative or JHSC to perform an inspection or regularly scheduled inspections.

Where necessary, a MOL inspector can apply to a Justice of the Peace or a provincial judge for a warrant specifically permitting an entry and search. The warrant can specify that the inspector be permitted to bring along an expert in a particular area (for example, an industrial hygienist who tests air quality) to assist in the inspection.

MOL inspectors typically come to the workplace for one of three reasons:

- to investigate an incident or alleged violation that has been reported to the MOL (for example, an accident, a work refusal, or an allegation, by a party, that the employer is breaking a safety law);
- to give advice to workplace parties about safety issues, often at their request (a "consultation"; the frequency of these has diminished in recent years, and workplace safety associations have begun doing more of this work); or
- to perform an unannounced spot check, known in the MOL as a "field visit."

The MOL relies on internal policies to help it decide which workplaces warrant unannounced inspections. In general, a workplace may be targeted for regular inspection if it has a history of accidents, worker complaints, or non-compliance with safety laws. Where an employer has demonstrated that it can manage health and safety well on its own—the employee injury rate is average or below average, there are up-to-date safety policies and procedures in place, and the employer has responded appropriately to past safety problems—the MOL will generally direct its enforcement attention elsewhere. From an employer's perspective, demonstrating an ability to be self-reliant from a health and safety perspective can mean operating with minimal government interference.

Offences and Penalties Under the OHSA

The OHSA creates penalties for non-compliance in part IX. Any person who is found guilty of not complying with a provision of the legislation, an order of a Director, or an order of the MOL is liable to pay a fine of up to $25,000, to serve a prison term of up to 12 months, or both. Lesser fines and/or prison terms are imposed in less serious cases.

For corporations that fail to comply, the maximum fine is $500,000. Again, lesser fines are imposed in less serious cases.

These penalties are imposed after a prosecution and a finding of guilt by the MOL; they are not simply handed out the moment an employer appears to be in non-compliance. While the OHSA provides for a penalty for failing to comply with a provision of the OHSA, in practice, a minor act of non-compliance by an employer that is rectified promptly upon being discovered by the MOL is unlikely to form the basis for a prosecution. However, where an employer's act of non-compliance is discovered because it causes a catastrophic accident with serious injuries or loss of life, a prosecution is the norm.

Regulations

Finally, part X of the OHSA gives the Lieutenant Governor in Council the authority to make regulations under the Act.

At the time this book was being written, 36 regulations were made under the OHSA. For a statute, this is considerably higher than average.

A regulation differs from a statute in that it is drafted by the executive branch of government and does not need to be passed into law by the parliament. The government uses regulations to communicate the details of how a statute should be administered. If every detail of its administration were put into a statute, some statutes would be unmanageably long and technical. Also, the government's ability to adjust minor details (measurements, formulas, standards, time periods, and so on) would be limited, because each minor change would require the passage of an amending statute. For this reason, much of this technical information is placed in regulations instead.

A list of the regulations under the OHSA as of spring 2008 appears in appendix A of this book.

Some of the regulations under the OHSA are quite brief, and address narrow subjects. For example, Training Programs (O. Reg. 780/94) says only one thing—that where a JHSC member needs to be certified (under OHSA section 26(1)(l)), the employer is required to "carry out" (and/or pay for) the certification training.

Some examples of the titles of other brief and narrow regulations are as follows:

- "Critical Injury—Defined";

- Joint Health and Safety Committees—Exemption from Requirements; and

- Teachers.

In addition, there are several individual regulations dealing with the handling of single hazardous substances, for example, arsenic, lead, and silica.

Several of the longer regulations deal with the OHSA's application to a particular industry. Examples of titles of these include:

- Diving Operations;

- Farming Operations;

- Health Care and Residential Facilities (a very long regulation, because of the wide range of hazards in health facilities); and

- Oil and Gas—Offshore.

If you have significant responsibility for occupational health and safety compliance in your organization, you should review the list of regulations carefully to determine if any of the industry-specific or risk-specific regulations apply to your workplace. Explaining the content of all of these regulations is beyond the scope of this book, though we will touch on the longest and most detailed regulation, Industrial Establishments, in chapter 6.

The Industrial Establishments regulation contains detailed administrative rules that apply to a wide range of workplaces. However, it does not apply to *all* workplaces. For example, it does not apply to hospitals, which are covered instead in the Health Care and Residential Facilities regulation. Be sure to check the list of regulations to determine which ones do and do not apply to your organization; you may need to supplement what you learn in this text with some independent study of regulations not covered here.

THE INTERNAL RESPONSIBILITY SYSTEM

Introduction

One of the key features of the OHSA is that it makes protecting worker safety a *shared* responsibility. It repudiates both the view that working (and earning) is a privilege that a worker assumes at her own risk, and the opposite view—that the employer is solely responsible for protecting workers. Instead, every workplace party is expected to participate in making the workplace safe. The net result of this expectation is known as the Internal Responsibility System (IRS).

The key participants in the IRS are:

- employers and constructors;
- supervisors (as defined by the OHSA, *not* as defined by the employer);
- the Health and Safety Representative or JHSC, and the Worker Trades Committee, if any;
- trade unions; and
- workers.

The OHSA also imposes safety-promotion responsibilities on certain other parties, including corporate officers and directors, owners of workplaces that are not construction projects, suppliers, licensees, and architects and engineers. The general responsibilities of these parties are summarized in the Facts and Figures feature below.

FACTS AND FIGURES

Duties of Other Workplace Parties

Besides the parties at the heart of the IRS, the OHSA also imposes duties, depending on the circumstances, on the following workplace parties:

1. Corporate Officers and Directors. Corporate officers and directors, as representatives of the employer, have a personal **duty of care** (reinforced by section 32 of the OHSA) to ensure that the corporation takes all reasonable steps to comply with the safety legislation, regulations, and policies to which the company is subject. Historically (especially before the passage of Bill C-45, mentioned in the previous chapter, which made it easier to find corporations guilty of offences), imposing criminal sanctions on an employer for safety failures required identifying specific corporate officers and directors who were responsible for accidents. Directors and officers who ignore compliance requirements remain vulnerable to both criminal sanctions and fines under the OHSA.

duty of care
in negligence law, an obligation on the part of one person to take into account the effect of his or her actions on another person

2. Owners. The OHSA defines an owner as:

> a trustee, receiver, mortgagee in possession, tenant, lessee, or occupier of any lands or premises used or to be used as a workplace, and a person who acts for or on behalf of an owner as an agent or delegate.

Besides having a general duty to ensure that a workplace complies with the OHSA, owners are responsible for ensuring that "prescribed facilities" are available and maintained. "Facilities" is a broad term, but could include, for example, a "quick-acting deluge shower" designed to wash hazardous materials off workers in an area where exposure is likely. "Prescribed" means required by law; for example, having a deluge shower in an area where skin contact hazards are used is prescribed by section 125 of the Industrial

Establishments regulation made under the OHSA. Owners may also, in some cases, need to seek MOL approval to make structural changes to a workplace, and may need to submit the relevant plans to the MOL for review.

3. **Suppliers.** The OHSA imposes duties on "suppliers," but makes it clear that a supplier is only someone who meets the narrow definition of supplying machines, tools, or equipment under a rental or lease (*not* a sale) arrangement (section 31(1)). This kind of supplier has a duty to ensure that the thing supplied is in good condition, complies with the OHSA, and is well-maintained.

4. **Licensees.** A licensee is also narrowly defined for the purpose of the OHSA. Under this statute, it means a party who holds a licence to cut Crown timber. These people or corporations are required to protect the health and safety of the workers who cut timber under the licence and to comply with the OHSA—specifically, with the logging-related rules made under it.

5. **Architects and Engineers.** An architect or engineer (defined as such by the legislation that governs his or her profession) can be liable, under the OHSA, for harm that comes to a worker (not necessarily an employee of the architect or engineer) because of incompetent advice given by the architect or engineer for the purpose of a certification required by the OHSA. For example, section 40 of the Industrial Establishments regulation requires that electrical equipment be certified by the Canadian Standards Association or the Electrical Safety Authority.

The following sections will describe the specific responsibilities of each IRS participant.

Employers

Employers have the greatest degree of control over the work environment, and therefore the largest number of duties under the IRS. Most important of all is a general duty, recognized under both the OHSA and common law, to take all reasonable precautions to protect the health and safety of their workers. Some of the more specific responsibilities of employers (and constructors, who are treated as employers under the legislation) are described in section 25 of the OHSA.

The duties of employers can be subdivided into two main categories: those that apply to all workplaces, and those that vary depending on the type of workplace.

The general and universal health and safety duties of employers can be grouped into the following categories:

1. **Duty of care.** All employers must:

    - take every precaution reasonable under the circumstances for the protection of a worker; and
    - assist in emergencies involving workers (which can mean providing any requested information, even confidential business information, to qualified medical professionals).

2. **Human resources management and training.** All employers must:

    - ensure that underage workers are not hired;
    - hire only competent people to act as supervisors;
    - provide information, instruction, and supervision to workers to help protect their health and safety;

- inform workers or their supervisors about hazards in the workplace;
- train workers in how to handle, store, use, dispose of, and transport any hazardous materials used in the workplace;
- post a copy of the OHSA and any relevant regulations in the workplace;
- post MOL information explaining the rights of workers under the OHSA (for example, the right to refuse unsafe work); and
- refrain from making reprisals against employees for exercising their rights under the OHSA.

3. **Maintenance of facilities and equipment.** All employers must:
   - ensure that the equipment, materials, and protective devices used in the workplace are maintained in good condition; and
   - ensure that every part of the physical structure of the workplace can bear the loads placed upon it;

4. **Occupational health and safety program.** All employers must:
   - develop a written health and safety policy (reviewed once a year) and design a program to implement it;
   - post a copy of the health and safety policy in the workplace;
   - oversee the appointment (by the workers) of a Health and Safety Representative or the establishment of a JHSC;
   - assist the Health and Safety Representative or JHSC in carrying out their functions;
   - provide the Health and Safety Representative or JHSC with any occupational health and safety reports that it has in its possession; and
   - advise workers of the results of any occupational health and safety reports.

5. **Post-incident compliance.** All employers must do the following in the situations described below:
   - Where any person is critically injured or killed at the workplace (the definition of "critical injury" is discussed in chapter 9 of this book), the employer must immediately notify an inspector at the MOL, the JHSC or Health and Safety Representative, and the union, if any.
   - When there is an accident, explosion, or fire at the workplace and a worker is disabled or needs medical attention, the employer must notify the JHSC or Health and Safety Representative, and the union, if any (in writing). In some cases, the employer may be required to notify the MOL.
   - When an employer learns that an employee (including a former employee) has an occupational illness or has made a **Workplace Safety and Insurance Board (WSIB)** claim based on one, the employer must notify an inspector at the MOL, the JHSC or Health and Safety Representative, and the union, if any.

Workplace Safety and Insurance Board (WSIB)
an Ontario tribunal charged with the mediation or adjudication of disputes arising under employment-related legislation, like the OHSA

Besides these universal duties, employers may have additional duties depending on what the legislation and/or regulations prescribe for a particular type of workplace. As you might imagine, workplaces that are more inherently dangerous are subject to greater occupational health and safety regulation than are workplaces with relatively few health or safety hazards. When the legislation refers to a duty to do something that is "prescribed," this means that there is another provision somewhere (in legislation or in a regulation) that explains what the employer must do. If you have significant responsibility for occupational health and safety compliance, you should review the list of prescribed duties carefully, noting the regulations that apply to your particular kind of workplace. This will allow you to determine which, if any, of these special duties apply to your employer.

Where prescribed by law, an employer must:

- provide and maintain any prescribed equipment, materials, or protective devices (the kind worn by workers are sometimes called personal protective equipment, or PPE), and ensure that these devices are used in accordance with the regulations;

- comply with the legislation and regulations governing biological, chemical, or physical agents. These rules usually include:

    a. maintaining records about the handling, storage, use, and disposal of these agents;

    b. notifying the MOL when one of these agents is used in or brought into the workplace;

    c. monitoring levels of these agents in the workplace and posting the results of this monitoring;

    d. complying with standards designed to limit worker exposure to these agents; and

    e. disclosing (via notices) worker exposure to these agents.

- report, to an inspector at the MOL, the JHSC or Health and Safety Representative, and the union, if any, accidents or unexpected events that could have caused injury (even if no one is hurt) when the accident or event occurs at a construction site, mine, or mining plant;

- establish and maintain an occupational health service for workers (this usually applies to larger employers in workplaces with well-known health and safety risks);

- create medical surveillance and safety-related medical testing programs. There are two kinds of medical surveillance programs: compulsory ones (for example, tuberculosis screening for hospital workers, designed to protect patients) and voluntary ones (designed to help a worker monitor his or her own personal health and safety). Where having a program is prescribed by law, an employer must pay for the tests and pay the employee for his or her time when the testing is being done;

- ensure that workers who have not been tested under a compulsory medical surveillance program (or who have failed the test or tested positive for something) are not allowed to work or be in the workplace; and

- provide written instructions to workers about how to protect themselves at work, and or carry out safety training programs for workers, supervisors, or members of the JHSC.

Many of these duties are discussed in more detail in chapters 8 and 9 of this book.

Supervisors

The OHSA assigns specific duties to supervisors. It is important to understand that these duties apply to supervisors *as defined by the OHSA*, which may not necessarily correspond with the employer's own definition of supervisor or the employer's scheme of job titles.

FACTS AND FIGURES

Who Is a Supervisor?

The OHSA defines a supervisor as "a person who has charge of a workplace or authority over a worker."

This definition has been interpreted by the **case law** to include not only those people who have the title "manager" or "supervisor," but also individuals who instruct and oversee workers in the use of equipment or materials and who have the authority to enforce employee compliance with their use of these things.

Workers who may fall into this category include, for example, forepeople, team leaders, or lead hands.

case law
the law as established by the outcome of former cases

As noted in the previous section, employers have a duty to hire only "competent" supervisors. These supervisors must in turn:

- ensure that workers work in a safe manner, taking every precaution reasonable under the circumstances for the protection of the workers;

- ensure that workers use or wear safety devices, equipment, and clothing properly, and adhere to the measures and procedures prescribed by the law or by policies of the employer;

- where prescribed (by law or a regulation), provide written instructions to workers about safety measures and procedures; and

- advise workers of potential or actual hazards in the workplace of which the supervisor is aware.

In order to carry out his duties under the OHSA, a supervisor must be reasonably knowledgeable about the following:

- what the known hazards in the workplace are;

- whether there are any unusual hazards at a given time (a supervisor is expected to monitor changes in the work area to determine this);

- which kinds of safety devices and equipment are required by the employer;
- the proper use of safety devices and equipment;
- the safety procedures that apply to the workplace; and
- the level of knowledge about safety procedures and equipment that each worker has (this is especially relevant, for example, when working with temporary workers or new hires).

Health and Safety Representative or Joint Health and Safety Committee

As noted earlier in this chapter, part II of the OHSA requires that workplaces appoint either a Health and Safety Representative (for small workplaces) or a JHSC. These IRS participants are therefore a creation of the legislation itself.

The central role of the Health and Safety Representative or JHSC is to monitor and manage occupational health and safety compliance in the workplace. For this reason, the duties of this participant are very detailed. A full discussion of Health and Safety Representatives and JHSCs appears later in this chapter under the heading "Health and Safety Representatives and Joint Health and Safety Committees."

Trade Unions

Not all workplaces are unionized. However, where there is a union in the workplace, the OHSA gives both rights and responsibilities to the union within the IRS.

As an organization devoted to representing the rights of workers, unions have traditionally played a key role in monitoring and promoting health and safety in the workplace. Unions are often at the forefront of new safety initiatives and policies, play a role in investigating workplace hazards and accidents, and support employees in their health and safety-related interactions with management.

FACTS AND FIGURES

What Is a Union for the Purpose of the OHSA?

The Ontario OHSA adopts the definition of "trade union" provided in the Ontario *Labour Relations Act, 1995* (OLRA). According to this, a trade union is

> an organization of employees formed for purposes that include the regulation of relations between employees and employers and includes a provincial, national, or international trade union, a certified council of trade unions and a designated or certified employee bargaining agency.

The OHSA union provisions apply only to a union that is recognized under the OLRA as being the exclusive bargaining agent for the affected workers.

The OHSA also counts as a union any organization that represents workers covered by the OHSA where that organization has exclusive bargaining rights under any other legislation in respect of those workers.

Figure 3.2 Union Notice Entitlements

Event/Circumstance	Notice Provider	Time Frame/Format
Death or critical injury to a worker	Employer	Immediately
Accident, explosion, or fire causing a non-critical injury resulting in disability or a need for medical attention	Employer	In writing, within four days
Employer has learned that a former member of the bargaining unit has an occupational disease	Employer	In writing, within four days
Statistics/data have been compiled about workplace safety performance/incidents	WSIB	Upon the union's request
A work refusal, at least where no worker JHSC member is available to represent the worker (so that the union can select a representative)	Employer	Unspecified, but the worker is entitled to immediate representation

Under the OHSA, unions play a key role in deciding who will be responsible for health and safety in the workplace. Where the workplace is unionized but is small enough to need only a Health and Safety Representative, the union is entitled to decide who will fill that post. Where a JHSC is required, the union will choose the worker members, will decide which ones will count as certified members (if more are eligible for this status than are required), and will often provide the training required for certification (at the employer's expense).

Where two or more workplaces apply for and are granted approval to share a multi-workplace JHSC (more on this below), the union(s) representing the workers will negotiate the terms of reference of this multi-workplace JHSC.

Unions play a key role in overseeing employer health and safety compliance. A significant part of this role amounts to seeing that a JHSC is properly constituted, that members are properly certified, and that the employer gives worker members paid time for JHSC activities. Unions will often assist employers in identifying workers to take over the duties of JHSC members so that those members can be free to discharge their duties (this is referred to, in some workplaces, as arranging "release time").

Where there is a lack of agreement, as between a workplace and the MOL, about whether a Health and Safety Representative or a JHSC is required, both the employer and the union are entitled to consult with the MOL on the issue.

Unions are entitled to notice, from various sources, of a number of different events. These notice entitlements are set out in figure 3.2.

Another area in which unions play a central role under the OHSA is in the context of worker allegations that an employer has made reprisals against a worker for exercising OHSA rights; as explained earlier, section 50 of the OHSA prohibits employers from making such reprisals. Where an employer is alleged to have done so, the worker is entitled to begin a legal proceeding before the OLRB in an effort to prevent or seek redress for the employer's actions (which may include demotions, firings, pay cuts, suspensions, denial of privileges, and so on).

Where there is no union, a worker seeking to challenge a reprisal must bring an "application" before the OLRB. Where the worker is represented by a union, he or she can usually choose between bringing a grievance under the **collective agreement** or an application under the OHSA (the union may seek to influence the worker's choice). Both an application and a grievance are heard by the OLRB, but the process is different—grievances are heard by an arbitration panel and are usually simpler, from a procedural standpoint. Applications are decided based on the provisions of the OHSA and the previous **jurisprudence** (cases) decided under it, whereas grievances are decided by interpreting the terms of the collective agreement that governs the worker. In both cases, the union plays a significant role in supporting the worker through the process.

Unions have standing to appeal orders made by the MOL, or to appeal the MOL's decision *not* to make an order. The appeal process is discussed in more detail in chapter 10.

collective agreement
a complex contract between a union and an employer that sets out most of the terms of employment that govern unionized employees

jurisprudence
the collective of legal decisions

LABOUR'S PERSPECTIVE ON OHS: DOES THE IRS GET THE EMPHASIS WRONG?

At its 2003 annual strategic planning conference, the Workers' Health and Safety Centre—Ontario's designated occupational health and safety training centre—reviewed a report that expressed some criticism, from a labour perspective, of the IRS created by the OHSA. The report suggested that a focus on shared responsibility for preventing workplace injuries and illnesses may distract from the more appropriate (in the opinion of the Centre) goal of eliminating workplace hazards—something only the employer has the ability to do:

> [M]any ... interpret IRS as 'individual' rather than internal responsibility, downloading upon workers key responsibilities for identifying and controlling workplace hazards, something over which the average worker has little control.
>
> This same mindset has also encouraged the proliferation of behaviour-based safety programs in our workplaces. These programs do little to further the real goal of health and safety which is to eliminate hazards and improve worker wellbeing. By wrongly focusing on worker behaviour it not only engenders a false sense of security, but it also means precious time and resources are diverted from the real task of rooting out and eliminating hazardous working conditions.

The report also quoted from an Ontario Federation of Labour report that recommended a stronger enforcement role for the MOL. The Ontario Federation of Labour was critical of the "dilution" of the MOL's enforcement role through its participation in consultation activities designed to support the IRS, and had the following to say:

> Labour supports the IRS as a *supplement* to external enforcement *but not as its substitute*. It is labour's position that the enforcement system should *drive* the IRS. The IRS should not drive the enforcement system [author's emphasis]. (Workers Health and Safety Centre Ontario, 2003, p. 2)

The report went on to cite accounts of workplaces in which the IRS, or at least the worker component of it, was essentially powerless to redress problems in the workplace. For example, workers were found to lack access to a JHSC, worker members were chosen by the employer rather than by the workers themselves, and discipline for reprisals against workers who raised safety issues were common. The report advocated for greater powers, on the part of workers, to compel the MOL to exercise its enforcement powers in support of the provisions of the OHSA.

Workers

Within the IRS, workers themselves play a key role in keeping their workplaces safe.

Workers have both rights and responsibilities. As discussed earlier in this chapter, workers have the right to stop or refuse work that they consider unsafe.

Besides the right to refuse unsafe work, workers have additional rights under the OHSA. These include the right to participate in health and safety initiatives by electing or appointing Health and Safety Representatives or JHSC members who will represent their interests vis-à-vis management; the right to be free from reprisals for raising issues of safety or for exercising other OHSA rights; and certain specific rights to information that can help them protect themselves from workplace hazards. The information to which workers have a right of access includes:

- a copy of the OHSA and the regulations made under it;
- a copy of the employer's occupational health and safety policy;
- information about hazards in the workplace;
- relevant reports on health and safety in the workplace;
- MOL inspection reports;
- reports and statistics about the workplace compiled by the WSIB; and
- information about hazardous materials in the workplace under the WHMIS (see chapter 4).

Along with these rights go responsibilities. Worker responsibilities under the IRS are summarized in figure 3.3.

The list of worker responsibilities, as expressed in figure 3.3, is deceptively brief; however, the essence of the list is that it is the responsibility of the workers them-

Figure 3.3 Worker Obligations

Work Safely	Use Protective Equipment	Report Hazards
• not use equipment or machinery, or perform work, in a way that might endanger any worker (including themselves) • refrain from rough-housing, running, boisterous activity, fighting, "pranks," "contests," or "feats of strength"—in other words, fooling around—in the workplace	• wear or use all safety equipment, devices, or clothing required by the employer • not tamper with, remove, or otherwise alter any safety equipment, devices, or clothing required by the employer	• report problems with safety equipment, devices, or clothing (including that items are missing) to a supervisor or the employer • report any observed hazard to a supervisor or the employer • report any observed violation of the OHSA or regulations to a supervisor or the employer

selves to work safely and to be aware of and take action to remedy workplace hazards. This is, obviously, a very onerous responsibility. If you think about it carefully, what are normally viewed as worker "rights" under the legislation—for example, the right to know about hazards and the right to refuse unsafe work—are actually essential requisites that make it possible for workers to discharge their duty to work safely. When considered in this light, it is easy to understand why the OHSA is so intolerant of employer reprisals for the exercise of safety "rights." Workers cannot effectively protect themselves (or, by extension, protect their employers' safety ratings, affordable WSIB premiums, productivity, and public image as a safe place to work) unless they are empowered to manage risks by accessing hazard information and by stopping work that exceeds an acceptable level of risk.

Thinking Safe

Employee Empowerment Councils

In his book *Working Safe: How to Help People Actively Care for Health and Safety*, E. Scott Geller describes an occupational health and safety innovation used by one of his consulting clients. The innovation is called an "employee empowerment council." The council is intended as a mechanism for placing the responsibility for safety enforcement in the hands of those for whom safety is the highest priority: the workers themselves. Under this system, an employee-based council hears evidence relating to safety incidents, and can make recommendations to both workers and management about how to avoid similar problems in the future.

Here is how such a council might work:

- Workers should elect or appoint representatives from among themselves to serve on the council for a particular term. (If the workplace employs several distinct trades, it would be useful to ensure that there is representation from at least most of these.)

- The council should have a manageable number of members that reflects the size of the workplace—a council of between 3 and 11 members is probably appropriate. (Having an odd instead of an even number of members makes it easier to have a clear majority in the case of voting.)

- The council might choose a chairperson and a secretary from among its members and any other members with special responsibilities, if useful.

- The council should come up with a definition for "reportable safety incident." Incidents falling within this definition should be made reportable to the council.

- The council should create procedures describing how it will collect evidence about the incidents reported to it. Evidence from both management and staff should be considered, and if there is a JHSC, evidence from the JHSC could be considered as well. The investigative process used by the council should be kept as simple and transparent as possible; it could involve simply having a meeting with any workers and supervisors involved in the incident.

> - After deliberating (talking among themselves in private), the council should come up with recommendations about how to avoid a similar incident in future. These recommendations might be directed at workers and/or management.
>
> - The recommendations should be put in writing, delivered to any relevant parties, and posted for the workers to read.
>
> - If management agrees, a policy could be developed requiring management to respond to the council's recommendations within a specific time period after receiving them.
>
> An employee empowerment council serves more than just the function of generating recommendations. It also promotes the view that workers should take ownership of their own safety and the safety of their colleagues, and acknowledges that workers are usually very well placed and competent to identify the root causes and possible solutions to safety problems.

HEALTH AND SAFETY REPRESENTATIVES AND JOINT HEALTH AND SAFETY COMMITTEES

Introduction

Health and Safety Representatives and JHSCs have already been mentioned several times in this chapter. However, as the key players in the IRS, they and their roles merit a more detailed discussion. As you have learned, smaller workplaces (generally those with between 6 and 19 regular employees) are required to appoint a Health and Safety Representative, while larger workplaces must create a JHSC.

Health and Safety Representatives

Health and Safety Representatives are described in section 8 of the OHSA.

A Health and Safety Representative serves as the liaison between the workers and the employer in relation to safety issues. It is a part-time role, taken on by one employee who discharges her health and safety duties in addition to her regular work responsibilities.

Very small workplaces (those with five or fewer regular employees) are exempt from having either a Health and Safety Representative or a JHSC. In these very small workplaces, there is generally close contact between the "employer" or his representatives and the "workers"; indeed, there may be little distinction between them. Safety is typically handled in a fairly open and informal way, and requiring a Health and Safety Representative would introduce a level of administrative complexity that would make little sense.

Where a Health and Safety Representative is required, this person should be chosen not by the employer, but by the workers. The only exception to the workers' right to choose is where the small workplace is unionized, in which case the union normally selects the representative. The Health and Safety Representative should not be a person who serves a managerial function, unless he or she is quite junior in the small company's hierarchy. The idea is to create some separation between this person and the employer so that the employees feel as though they are being represented by a peer.

Health and Safety Representatives need no special training or certification. They must be granted time away from their regular duties to attend to health and safety matters, and they must be paid for their time.

Health and Safety Representatives have fewer duties than JHSCs (for example, they need not hold meetings); however, like JHSCs, they are responsible for identifying hazards in the workplace and for bringing those hazards—along with recommendations for how to improve occupational health and safety—to the employer.

Joint Health and Safety Committees

JHSCs are required in all workplaces with 20 or more regular employees, on construction projects expected to last at least three months, and in all workplaces where a designated substance regulation applies.

COMPOSITION OF THE COMMITTEE

A JHSC is made up of members from both the general workforce and from the employer/management. The size of the JHSC required varies depending on the size of the workplace.

Where there are fewer than 50 workers, the *minimum* number of members on the JHSC is two: one from the worker side and one from the management side.

Where there are 50 or more workers, the *minimum* number of committee members is four: two from the worker side and two from the management side.

There is no maximum number of members on a JHSC. However, a committee's management members cannot outnumber the worker members; in other words, at least half of the members must be from the worker side. Worker members *can* outnumber management members. The MOL makes it clear that there should be enough members on a JHSC to provide meaningful representation for all aspects of the workforce:

> Whenever possible, committees should represent the health and safety concerns of the entire workplace. For example, if a workplace has a plant, office, laboratory and warehouse, each of these areas should be represented on the committee. (Ontario Ministry of Labour, 2002b)

The authors of the *HR Manager's Guide to Health and Safety* recommend that, in choosing the management members, the employer should ensure that at least one of them is a "person of influence"—a manager who is perceived to have real influence on company policy. This will increase the effectiveness of the JHSC and give the workers confidence that the committee has the ear of management.

Sometimes, it may not be entirely clear whether a particular person who is interested in serving on the JHSC would qualify as a worker member or a management member. The MOL provides this guidance on the issue:

> An employee who has the authority to discipline, hire, fire or recommend discipline, hiring or firing is considered a managerial employee. He or she can serve on the committee, but not as a worker representative. (Ontario Ministry of Labour, 2002a, p. 19)

> # HARD LESSONS
> ## Protecting the Appearance of Equality
>
> The OHSA places no upper limit on the number of members a JHSC can have; however, it *does* require that the number of management members not exceed the number of worker members.
>
> While the reason for this prohibition is not explicitly set out in the Act, the rule is in keeping with the OHSA's general promotion of a sense of shared ownership of the health and safety function between management and workers. Having non-management workers form at least half of the JHSC sends the message that health and safety decisions in the workplace will not be driven by the employer's own personal agenda.
>
> In 2005 the OLRB considered a situation in which a number of management-level employees who were not JHSC members were sitting in on JHSC meetings (*Minto Developments Inc. v. Duquette*). The presence of these individuals meant that management regularly outnumbered workers at the meetings.
>
> Upon an inspection of the workplace, an MOL inspector reviewed the minutes of the JHSC meetings and learned of the management "spectators." Along with eight other orders issued against the employer, the inspector ordered that the extra management staff be excluded from JHSC meetings.
>
> The employer attempted, under section 61(7) of the OHSA, to have this order suspended pending a hearing on its validity. The employer contended that it was useful to the business of the JHSC to have knowledgeable people in attendance at the meetings, even though these people were not JHSC members. The OLRB disagreed.
>
> When asked to suspend an order, the OLRB must consider three questions:
>
> 1. whether granting the suspension is not likely to endanger worker safety;
> 2. whether the person who is bringing the appeal (in this case, the employer) will suffer greater "prejudice" (usually, harm or disadvantage) if the order is allowed to stand than the person being protected (in this case, the workers) would suffer if the order were suspended; and
> 3. whether the person bringing the appeal can establish a *prima facie* case that the order will be successfully appealed (meaning that the evidence supporting the appellant's wish to have the order overturned is convincing at first glance).
>
> The OLRB concluded that it could not answer any of the above questions with a "yes." For that reason, the order remained in place.
>
> This decision provided little in the way of explicit detail about the inspector's concerns. It is reasonable to infer, however, that the inspector believed that the presence of extra management employees at the JHSC meetings might prove intimidating to workers or might have an adverse effect on the worker members' confidence when it came to speaking out about their concerns. The case suggests that not only must management–worker equality be demonstrated in the membership of a JHSC, but that the *impression* of equality must be maintained. While the employer was willing to allow management employees to sit in on the JHSC sessions, it is unlikely that workers had the same kind of freedom to leave their regular duties to do the same.

While the OHSA specifies no particular term for committee membership, the MOL recommends that members serve terms of at least one year in order to promote continuity in JHSC knowledge and performance. Ideally, members' terms should be staggered, so that the replacement of multiple members at once (with a resulting loss of knowledge and experience) can be avoided.

CERTIFICATION OF COMMITTEE MEMBERS

In permanent workplaces with 20 or more workers, and on construction projects employing at least 50 workers, the OHSA requires that at least one JHSC member from each side be certified.

A certified member is a member who has received prescribed training in health and safety law and in the identification of workplace hazards. Standards for certification training are set by the WSIB (visit their website at www.wsib.on.ca). Some of the actual training is provided by providers approved for this purpose by the WSIB; other parts of the training are provided by sector-specific health and safety associations (a list of these appears in appendix B of this book) or by private or in-house training providers. The employer's responsibilities with respect to certification of JHSC members are described in more detail in chapter 8.

MULTI-WORKPLACE JHSCs

Sometimes, a single employer operates more than one workplace with 20 or more workers. To reduce the duplication and administrative complexity that might flow from the employer having to interact with multiple JHSCs, the MOL has a policy of allowing these employers to request approval of a single JHSC to govern more than one workplace. Where the MOL grants its approval, the JHSC thus created is known as a multi-workplace JHSC.

In general, multi-workplace JHSCs have more members than the minimum (four) stipulated for large workplaces. Members representing each workplace should take part. Where there is more than one trade union representing workers covered by the JHSC, each union will have at least one JHSC member to represent it.

One of the key challenges of a multi-workplace JHSC is ensuring that workers have ready access to their JHSC and feel that the JHSC is in touch with the concerns at each workplace. It will be up to the members of the multi-workplace JHSC (and to the worker-side members in particular) to maintain a visible and accessible profile within the individual workplaces, so that workers feel that the JHSC is there for them when they need it. Section 9(32) of the OHSA provides that the employer must post the names and work locations of all committee members in a location readily visible to the workers. Where there is a multi-workplace JHSC, it would be a good idea to post contact information for members (for example, phone numbers and email addresses) in this location as well.

Mandate and Duties

Section 9(18) of the OHSA sets out the powers and the function of a JHSC. A JHSC

- identifies workplace hazards (including by hearing employee complaints and concerns);
- evaluates these hazards;
- makes recommendations about how to improve workplace safety to the employer (or constructor) and to the workforce; and
- follows up on and monitors the employer's establishment of safety programs, measures, and procedures.

In order to properly carry out these functions, a JHSC has a right

- to obtain information from the employer or constructor about "the identification of potential or existing hazards of materials, processes or equipment," and "health and safety experience and work practices and standards in similar or other industries of which the constructor or employer has knowledge"; and

- to obtain information from the employer about any health and safety-related testing the employer performs or has performed on any "equipment, machine, device, article, thing, material or biological, chemical or physical agent" in the workplace, and—where appropriate—to be consulted about, and/or to monitor (via the presence of a worker-side JHSC member), any such testing.

Procedures

MEETINGS

The OHSA provides that the JHSC should meet at the workplace at least once every three months. The employer is required to provide a meeting place and to give members time away from their regular duties to prepare for and to attend meetings. This time must be paid, either at the members' regular rate or at any premium rate that may apply (for example, overtime or a night shift premium).

JHSC meetings are co-chaired by two members, one from each of the management and worker sides.

Minutes of the JHSC meetings must be recorded (usually by a person chosen to act as secretary) and must be made available to a MOL inspector on his or her request.

The purpose of JHSC meetings is general, and the business that must be done at each meeting will vary depending upon what has been going on at the workplace. Members may need to plan inspections, review the results of previous inspections, discuss the results of investigations, prepare recommendations to the employer, prepare reports for the MOL, or plan training or other health and safety initiatives, among other tasks.

The MOL notes that its inspectors can attend JHSC meetings at workplaces at the request of JHSC members.

INSPECTIONS

As noted above, one of the key functions of the JHSC (or Health and Safety Representative, in a smaller workplace) is to identify hazards. In some cases, workers report hazards to the JHSC; in other cases, hazards come to light in the course of work refusals or stoppages. However, it is not enough for a JHSC to sit idly by waiting to hear about hazards. JHSC members must actively educate themselves about hazards in the workplace by conducting inspections on a regular basis, at least once a month (in very large workplaces, this schedule can be altered so that work areas are inspected on a rotating basis).

Where there are certified members on a JHSC, these members should be chosen to conduct the inspections. There are many sources of information on how to conduct inspections; chapter 8 discusses these in more detail.

HANDLING WORKER COMPLAINTS AND WORK REFUSALS

When a worker has a concern or complaint about occupational health and safety in the workplace, the worker is expected to report it first to his or her supervisor (OHSA section 28). However, if the supervisor does not resolve the problem to the worker's satisfaction, the worker (or the supervisor) can bring his or her concern to a JHSC member.

The JHSC can then make a recommendation to the supervisor about how to resolve the problem. If the problem is still not resolved, the JHSC can contact the MOL for advice.

When a worker or the employer reports a work refusal or stoppage to the JHSC, the JHSC is required to take part in the investigation of the circumstances of the refusal or stoppage (the investigation is generally performed by a supervisor). If the work stoppage results in a call to the MOL, the JHSC is entitled to have a member present during the MOL's investigation of the matter. (In workplaces with just a Health and Safety Representative, that representative assumes this role.)

FACTS AND FIGURES

Protecting the Confidentiality of Personal Information

While conducting inspections or investigations, or in the course of addressing worker concerns and complaints, a JHSC may receive confidential personal information about individuals. The appropriate handling of this personal information is very important.

Personal privacy is carefully protected under Canadian law. A full discussion of privacy law is beyond the scope of this book, but resources on privacy protection exist in print and on the Internet. Most employers have (or should have) a written privacy policy, and the Health and Safety Representative or JHSC should be given a copy of this policy to study and to follow.

In general, personal information received for a particular purpose must be used only for that purpose, and only by the person who receives it. If a JHSC member wishes to share the information with someone else (a MOL inspector, for example, or a health care practitioner), the person who has provided the information must be asked for permission—every time. In some cases, providers of personal information can be asked to sign carefully drafted releases or waivers to simplify the handling of their information. The employer's waiver procedures should be incorporated into the workplace's privacy policy.

The need to protect confidential information also extends to information about the employer—for example, test results or trade secrets. Permission from the employer must be obtained any time the JHSC or Health or Safety Representative wishes to disclose this kind of information to a third party.

SERIOUS ACCIDENT INVESTIGATIONS

Any time a person is killed or critically injured in a workplace incident, the JHSC is required to investigate the incident. This investigation is conducted by worker members of the JHSC, who are entitled to inspect the accident scene as well as any relevant equipment or devices. The members who conduct the investigation are required to report their findings to the JHSC as a whole. After this is done, the JHSC is expected to make recommendations to the employer (and/or the workforce) about how similar accidents can be avoided in the future. The JHSC should then receive the employer's response to the recommendations within 21 days, and should follow up on the implementation of any suggested programs, practices, or other solutions that were the subject of the recommendations or that were described in the employer's response.

There are resources available to assist JHSC members in conducting investigations and preparing recommendations; chapter 8 discusses these in more detail.

The employer or a worker should contact the MOL for guidance if a dispute arises within the workplace about:

- whether a Health and Safety Representative or a JHSC is required;
- how many members are required on the JHSC;
- how many or whether any members need to be certified;
- the employer's provision of time or pay (or other support) to the JHSC; or
- any aspect of the functioning of a JHSC.

The MOL may be able to provide useful advice, or, where advice fails to resolve the issue, a Director at the MOL has the power to issue an order requiring the creation of a JHSC or changes to its operation.

KEY TERMS

Internal Responsibility System (IRS)

Health and Safety Representative

Joint Health and Safety Committee (JHSC)

Workplace Hazardous Materials Information System (WHMIS)

reprisal

warrant

duty of care

Workplace Safety and Insurance Board (WSIB)

case law

collective agreement

jurisprudence

REVIEW QUESTIONS

1. "The OHSA is enforced by both an external responsibility system and an internal responsibility system." Explain.

2. Which system regulates the handling of hazardous materials in the workplace? Does the OHSA apply to hazardous materials or substances, and if so, how?

3. Which kinds of workers are *not* allowed to refuse unsafe work?

4. What kind of content is covered in the regulations made under the OHSA? Who enforces these regulations? Are there penalties for failure to comply with the regulations?

5. Where, besides in the OHSA itself, would you look for employer occupational health and safety responsibilities?

6. Why do you think the OHSA establishes different duties for employers and for supervisors, who are, in a sense, the agents of employers?

7. List three rights of workers under the OHSA. List three workers' duties.

8. Why are trade unions given a role within the IRS?

9. Who chooses a Health and Safety Representative for a workplace? Are there any restrictions on who may be chosen?

10. How many worker members must a JHSC have?

11. How do Health and Safety Representatives and JHSC members identify hazards in a workplace?

12. What should workers do if they are unhappy with the way the JHSC is representing their health and safety interests?

AT ISSUE: IS TOTAL INJURY PREVENTION A REALISTIC GOAL?

The authors of the *HR Manager's Guide to Health and Safety* suggest that an IRS should reflect the following three principles:

- **Safety First and Safety Always**: "[T]he central principle here is that a job that cannot be done safely should not be done."

- **Safety is Everyone's Responsibility**: "This is not simply a statement of good intent. It is a legal responsibility."

- **Safe Work is Efficient Work**: "Accidents should be considered an unacceptable cost of production. This view is legally and morally sound. It is also good business practice." (Knight, McCreadie, and Winch, 2004, pp. xii–xiii)

However, E. Scott Geller suggests that there are inherent dangers in promoting the idea that all accidents are preventable (Geller, 1996). These dangers include:

- the fact that, if all accidents are considered preventable, investigators may feel pressure to zero in on a single cause (or at-fault individual) and may overlook the fact that many accidents have multiple causes; and

- the possibility that, if employees and employers are pressured to reduce the accident rate to zero, and if they are offered rewards for doing so, injured workers and/or supervisors of injured workers may be pressured not to report accidents (which eliminates the opportunity to uncover causes or come up with prevention strategies).

1. Do you believe all injuries are preventable? Why or why not?

2. If you were an employer, do you think it would be better to reflect, in your health and safety communications to employees, a commitment to the elimination of all workplace injuries and illnesses, or to acknowledge the possibility that some injuries and illnesses are not preventable?

3. What kinds of safety promotion strategies risk provoking non-reporting of accidents?

4. How might you promote employee commitment to workplace accident prevention while encouraging reporting of workplace injuries?

CHAPTER 4
Introduction to the WHMIS

CHAPTER OBJECTIVES

After completing this chapter, you should be able to:

- Describe the legislative basis for the Workplace Hazardous Materials Information System (WHMIS).
- Explain the goals of the WHMIS.
- Summarize the components of the WHMIS.
- Describe the process for classifying products governed by the WHMIS.
- Explain the key compliance requirements for suppliers and importers of hazardous products.
- Explain the key WHMIS compliance requirements for employers.

INTRODUCTION

The Workplace Hazardous Materials Information System (WHMIS) was introduced in 1988. It is a national system developed as a cooperative effort by the federal, provincial, and territorial governments. The legislative underpinnings of the system will be discussed in the next section.

The WHMIS is being discussed in this book because compliance with the system is an important aspect of promoting occupational health and safety. It is impossible to be in compliance with the OHSA without complying with the WHMIS, because the latter establishes national standards for the handling of hazardous materials destined for or used in the workplace.

The main goal of the WHMIS is to establish national standards for measuring, classifying, and regulating hazardous materials. The system achieves this by prescribing activities such as:

- analysis of materials;
- classification of materials;
- labelling of materials;

- distribution of information to users (often employees); and
- user training.

The WHMIS regulates the activities of two main parties: suppliers of hazardous materials (who must provide product information and appropriate labelling before distributing or importing hazardous materials) and employers (who must provide information to employees about the properties of hazardous materials and how to use them safely).

From a business perspective, the details of WHMIS compliance will depend on the nature of the business. If a company is a supplier of a WHMIS-controlled product, it will need to comply with federal law that imposes the obligation to develop communications materials (labels and information sheets) that accurately describe the composition and properties of the product. If a company is an importer of WHMIS-controlled products, it will need (also under federal law) either to obtain or develop safety information for the product, and arrange for labelling. Finally, if a company is a user of WHMIS-controlled products, it will need to comply with provincial legislation that prescribes duties for employers.

Of course, in many cases, an employer will need to address compliance of more than one type—a business may both supply and use, or import and use, WHMIS-controlled products.

In the following sections, you will learn about how the WHMIS was developed and about the legal framework that supports it; about the goals of the system and how products are classified under it; and what is required for compliance on the part of suppliers, importers, and employers.

OVERVIEW OF THE SYSTEM

History

As you learned in the Introduction, the WHMIS was introduced in 1988 after a consultation process that involved the federal, provincial, and territorial governments. The federal provisions were implemented first, with the provinces following suit. The WHMIS is now fully implemented across Canada, and following the system's established rules is an essential element of occupational health and safety compliance for employers. The system has been generally well accepted by all participants, and Health Canada offers a possible explanation for its success (on its website at www.hc-sc.gc.ca):

> WHMIS represents an excellent example of synchronization and cooperation amongst the federal, provincial and territorial governments. This coordinated approach avoided duplication, inefficiency through loss of scale and the interprovincial trade barriers that would have been created had each province and territory established its own hazard communication system. The federal *Hazardous Products Act* established the national standard for chemical classification and hazard communication in Canada and is the foundation for the workers' "right-to-know" legislation enacted in every province and territory. A single national system for a mobile labour force also minimized confusion and facilitated implementation of the system.

Legislative Framework

As explained above, the WHMIS has both federal and provincial legislative components.

FEDERAL ASPECTS

The federal statutes that play a role in supporting the WHMIS, and the regulations made under them, are as follows:

Statute: The *Hazardous Products Act* (HPA)
Regulation: The Controlled Products regulations (CPR)
Statute: The *Hazardous Materials Information Review Act* (HMIRA)
Regulation: The Hazardous Materials Information Review regulations (HMIRR)

The provisions of the HPA are aimed primarily at suppliers (and importers) of hazardous products. The Act imposes two key requirements on suppliers—standardized labelling, and the creation of product information sheets (called Material Safety Data Sheets)—as a condition of any sale of WHMIS-controlled products destined for use in a Canadian workplace.

Schedule I of the HPA provides a *non-exhaustive* list of products that are hazardous and controlled under the WHMIS, for suppliers' basic reference.

The CPR explains how suppliers should determine whether the WHMIS applies to a particular product, and the classification of that product.

One of the benefits of having products classified under federal rather than provincial legislation is that this results in national standards, as Health Canada explains:

> The *Controlled Products Regulations* establish a *national* standard for the classification of hazardous workplace materials. In addition to setting out criteria for biohazards, chemical and acute hazards, the regulations specify criteria for chronic health hazards including mutagenicity, carcinogenicity, embryo and reproductive toxicity, respiratory tract and skin sensitization.

The HMIRA creates an administrative agency called the Hazardous Materials Information Review Commission (HMIRC). The commission rules on applications made by suppliers who are seeking exemption from certain WHMIS rules in respect of certain products. According to Health Canada, "WHMIS balances workers' right-to-know with industry's right to protect confidential business information." The WHMIS recognizes that, in some cases, the formulation of a hazardous product is confidential business information (a trade secret), and that the requirement to disclose ingredients could harm the business interests of a supplier. Together with the HMIRR, the HMIRA creates a procedure allowing suppliers to apply to withhold certain information from a product label or **Material Safety Data Sheet (MSDS)**.

PROVINCIAL ASPECTS

When the WHMIS was created, each province and territory was given the task of creating its own WHMIS provisions. In Ontario, the provincial aspects of the WHMIS are prescribed under the OHSA and by the Workplace Hazardous Materials Information System (WHMIS) regulation made under the OHSA.

Material Safety Data Sheet (MSDS)
a document made available in the workplace that is designed to describe, to workers, the risks associated with a hazardous material and the appropriate safety precautions that should be followed in handling the material

The provincial provisions prescribe the WHMIS rules that govern employers. The WHMIS regulation under the OHSA regulates certain features of the product designation system, labelling requirements, and MSDS requirements by employers. It also imposes training requirements that are not a part of the federal WHMIS rules.

As you will see below, the labelling and MSDS requirements for employers are somewhat different from those imposed on suppliers; however, the way in which products are classified conforms to the national standard.

HARD LESSONS

Reprisals for Employees' WHMIS Inquiries Lead to Damage Awards

Facts: Shortly after beginning work at an auto parts plant, an employee began feeling ill while working with chemicals used in the plant. She raised her concerns with the plant management, and asked specifically about the availability of WHMIS training, MSDSs for the chemicals in the plant, and the availability of personal protective equipment. She received no response from management.

As a result of the management's failure to address the employee's concerns, three other employees contacted the MOL for help. The Ministry sent an inspector to the plant and the inspector issued orders against the employer.

The employer expressed displeasure with the employees' actions and advised them to speak to management—and not the Ministry—the next time they had concerns. Two weeks after the employees' phone call to the Ministry, the employer scheduled a WHMIS training session; when the employees arrived, they were immediately laid off. According to the employer, the reason for the dismissal was "because of everything that happened."

Arguments: Two of the employees applied to the OLRB for damages for lost wages, alleging that the layoffs constituted reprisals. The employer did not make submissions at the hearing, but since the Board had denied its request for an adjournment, the hearing proceeded anyway.

Decision: The OLRB found that the employer's actions in laying off the employees because of their call to the Ministry amounted to reprisal under section 50 of the OHSA. It awarded the employees an amount representing wages lost between the date of the layoff and the date of the decision (minus an amount deducted for the employees' mitigation of their losses—that is, for pay they had received from other sources in the interim). One employee was awarded $7,584.00, while the other received $3,617.00. The Board also ordered that the employer reinstate the employees; however, the employees chose not to return.

Business Lesson: Even new employees are entitled to WHMIS training and to view product MSDSs. Having operated a shop without WHMIS compliance in the past should not lull employers into continued neglect of WHMIS responsibilities, because employees are entitled to seek help from the Ministry if their concerns are not adequately addressed in the workplace. And of course, reprisals for the exercise of OHSA or WHMIS rights will never be tolerated.

Zaluski v. Valthane Inc., 2000 CanLII 3201 (OLRB).

CLASSIFICATION OF HAZARDOUS MATERIALS

Any product is a WHMIS-controlled product if it meets one or more of the hazard criteria described in the Controlled Product regulations (CPR) and is not exempt from MSDS and labelling requirements under section 12 of the *Hazardous Products Act* (HPA).

Section 12 of the HPA provides:

> This Part [that is, the labelling and MSDS requirements] does not apply in respect of the sale or importation of any
>
> (a) explosive within the meaning of the *Explosives Act*;
>
> (b) cosmetic, device, drug or food within the meaning of the *Food and Drugs Act*;
>
> (c) control product within the meaning of the *Pest Control Products Act*;
>
> (d) nuclear substance, within the meaning of the *Nuclear Safety and Control Act*, that is radioactive;
>
> (e) hazardous waste;
>
> (f) product, material or substance included in Part II of Schedule I and packaged as a consumer product;
>
> (g) wood or product made of wood;
>
> (h) tobacco or a tobacco product as defined in section 2 of the *Tobacco Act*; or
>
> (i) manufactured article.

As you may have noticed in the above provision, certain products, while hazardous, are exempt from the application of the WHMIS because they are regulated under other legislative schemes (like the *Food and Drugs Act* or the *Explosives Act*) that provide their own systems for safeguarding users.

The hazard criteria that trigger the application of WHMIS duties are set out in part IV of the CPR. Part IV begins with a list of defined terms, at section 32. The sections that describe hazard criteria are sections 34–66. An outline of the various hazard classifications and subclassifications appears in figure 4.1.

The sections themselves provide specific descriptions of the scientific measures and findings that support these various classifications. Because of the highly technical nature of their content, the specific classification provisions will not be covered here. Suppliers and importers will need to seek the advice of scientifically trained personnel to assist in the process of classification and WHMIS compliance. Supplier and importer compliance is discussed in the next section.

SUPPLIER AND IMPORTER COMPLIANCE REQUIREMENTS

Introduction

Supplier and importer compliance with the WHMIS is prescribed by the Controlled Products regulations (CPR) made under the (federal) *Hazardous Products Act* (HPA). Employer compliance under the same WHMIS regime is governed by a *provincial* statute: the OHSA. Suppliers who are also employers will need to comply with both

Figure 4.1 WHMIS Hazard Classifications

CLASS A—COMPRESSED GAS

CLASS B—FLAMMABLE AND COMBUSTIBLE MATERIAL

> *Division 1: Flammable Gases*
> *Division 2: Flammable Liquids*
> *Division 3: Combustible Liquids*
> *Division 4: Flammable Solids*
> *Division 5: Flammable Aerosols*
> *Division 6: Reactive Flammable Materials*

CLASS C—OXIDIZING MATERIAL

CLASS D—POISONOUS AND INFECTIOUS MATERIAL

> *Division 1: Materials Causing Immediate and Serious Toxic Effects*
>> Subdivision A: Very Toxic Material
>> (causing "acute lethality")
>> Subdivision B: Toxic Material
>> (causing "acute lethality")
>
> *Division 2: Materials Causing Other Toxic Effects*
>> Subdivision A: Very Toxic Material
>> (causing chronic toxic effects, teratogenicity and embyotoxicity (harm to unborn babies), carcinogenicity (tending to cause cancer), reproductive toxicity (tending to impair fertility), respiratory tract sensitization, or mutagenicity (tending to cause cellular changes))
>> Subdivision B: Toxic Material
>> (causing chronic toxic effects, skin or eye irritation, skin sensitization, or mutagenicity)
>
> *Division 3: Biohazardous Infectious Material*

CLASS E—CORROSIVE MATERIAL

CLASS F—DANGEROUSLY REACTIVE MATERIAL

statutory systems; fortunately, the WHMIS provisions of each are harmonized, so there is no conflict. In general, the federal provisions impose more stringent disclosure requirements on packaged products, while the provincial rules require labelling of a wider range of vessels (including, for example, industrial piping that carries a hazardous product) and the provision of user training.

Classification

The first requirement imposed on suppliers and importers is the requirement to determine whether a material destined for use in the workplace is a WHMIS-controlled product.

As explained above, the hazard criteria are set out in part IV of the CPR, beginning at section 32 of the regulations. The information in this part of the CPR is highly technical, and suppliers and importers should obtain the help of staff with a scientific background to interpret the criteria.

Determining whether a product is a WHMIS-controlled product—and if so, which class—can be a complicated exercise. According to Health Canada (on their website), suppliers and importers are *not* legally required to undertake tests to determine the WHMIS classification of a product:

> As a fundamental principle, during the development of WHMIS, all stakeholders agreed that nothing in the hazard criteria, nor any part of WHMIS, would require additional toxicological testing. Rather, WHMIS was designed to make the best use of existing toxicological data.

The one exception to this rule is that an importer or supplier of a WHMIS-controlled product who is obliged to prepare an MSDS and labelling for the product but does not have adequate information about the product ingredients may, if it cannot obtain the necessary information, need to test the product to determine its composition so that it can comply with disclosure requirements.

In general, however, in making determinations about the application of the WHMIS or about product classification, suppliers and importers are expected to make use of information already available about the product. This information can come in any form, including:

- ingredient lists from the manufacturer;

- information from the manufacturer about the product's properties and associated hazards;

- published results from scientific studies of the product's effects;

- hazard information available about other products that are very similar to the product; and

- anecdotal information that the supplier or importer may have gained through using the product (for example, complaints from users about a harmful effect).

Section 33 of the CPR establishes the procedure for classifying hazardous products. In summary, in order to determine the classification of a product, a supplier will:

1. Have reference to results from tests that the supplier *itself* has conducted on the product;

2. Evaluate and use scientific judgment in considering test results on the same product conducted by others; or

3. Evaluate and use scientific judgment in considering the relevance to the supplier's product of test results on a *different product that has similar properties* that have been conducted by others.

The section then provides (at subsection 2) that, in making a determination about whether a product is Class D (poisonous and infectious material), the supplier may "use information of which the supplier is aware or ought reasonably to be aware" concerning the product.

Subsection 3 of section 33 goes on to explain that the test results a supplier relies upon in subsections 1 or 2 must have been obtained by tests that meet recognized scientific standards, which are then described in detail. For example, at section 33(3)(ii):

in the case of a test for skin or eye irritation, the Draize Test as described in volume 82 of *The Journal of Pharmacology and Experimental Therapeutics*, dated 1944, at pages 377 to 390.

To assist suppliers in exercising the discretion (that is, the "evaluation" and "scientific judgment") implied in section 33, a technical subcommittee of the WHMIS Current Issues Committee (CIC) was formed. The subcommittee developed guidelines for the use of professional judgment in making classification decisions; the guidelines are reproduced in appendix C. They provide, for example, that for the purpose of section 33(2):

> Every supplier "ought reasonably to be aware" of appropriate published literature. The Canadian Centre for Occupational Health and Safety (CCOHS) is one organization capable of conducting a comprehensive literature search. When additional information is made available to the supplier by the appropriate regulatory agencies, by industry or trade association(s), and by labour organization(s), the supplier is expected to evaluate that information.

Also, Health Canada makes it clear that there is no penalty for overclassifying a product (classifying it as more dangerous than it might actually prove to be) in an excess of caution.

Like suppliers, importers are subject to classification requirements. Before importing a product destined for Canadian workplaces, an importer must either obtain or prepare an MSDS for the product, and must ensure that the product bears a label that complies with the hazard disclosure standards of the WHMIS (discussed below). If an accurate MSDS and label are not available from the import source, the employer must take the same steps to classify the product (from section 33) that a supplier who produces the product here in Canada would. This must be done *before* importation. (To make this easier, the law creates an exemption for the importation of laboratory samples.)

Once a product has been evaluated and classified, its classification will dictate certain disclosure requirements as required by the WHMIS.

Disclosure by Suppliers and Importers

Before an importer can bring a WHMIS-controlled product into Canada for eventual workplace use or before a supplier can supply such a product to a workplace, the importer or supplier must develop and make available two key disclosure documents.

The first "document" is a product label. Each individual packaged unit of the product must have a label applied to it that discloses hazards associated with the product. The format and content of the label must conform to the specific requirements of the WHMIS, which are explained in the CPR.

A label can come in a variety of forms. It can be a sticker applied to a container, a painted or inked message printed or stamped directly on a container, a tag affixed securely to a container (there are various ways to affix a WHMIS label; the appropriate method will be dictated by the size and shape of the container). Where a container is very small (contents less than 100 millilitres), a supplier is permitted to use an abbreviated form of label; otherwise, the full label is required.

Some key features of a WHMIS-compliant label are as follows:

1. Hatched Border. WHMIS labels feature a distinctive hatched (a line with diagonal line breaks) border.

2. Hazard Symbols. Labels must include the corresponding WHMIS symbol for each of the hazard classifications into which the product falls. The WHMIS hazard symbols and their meanings are shown in figure 4.2.

3. Printed Information. Labels must contain, in both English and French, the following information:

    - the name of the product or (if the name is protected as a trade secret) an identifying code name or number;

    - the name of the supplier or importer;

    - a statement that there is an MSDS available for the product;

    - a risk phrase or phrases describing the dangerous properties of the product (for example, "poison" or "skin irritant");

Figure 4.2 WHMIS Labels

	Class A: Compressed Gas	Contents under high pressure. Cylinder may explode or burst when heated, dropped, or damaged.
	Class B: Flammable and Combustible Material	May catch fire when exposed to heat, spark, or flame. May burst into flames.
	Class C: Oxidizing Material	May cause fire or explosion when in contact with wood, fuels, or other combustible material.
	Class D, Division 1 Poisonous and Infectious Material: Immediate and serious toxic effects	Poisonous substance. A single exposure may be fatal or cause serious or permanent damage to health.
	Class D, Division 2 Poisonous and Infectious Material: Other toxic effects	Poisonous substance. May cause irritation. Repeated exposure may cause cancer, birth defects, or other permanent damage.
	Class D, Division 3 Poisonous and Infectious Material: Biohazardous infectious materials	May cause disease or serious illness. Drastic exposures may result in death.
	Class E: Corrosive Material	Can cause burns to eyes, skin, or respiratory system.
	Class F: Dangerously Reactive Material	May react violently causing explosion, fire, or release of toxic gases, when exposed to light, heat, vibration, or extreme temperatures.

Source: Canadian Centre for Occupational Health and Safety, "OSH Answers: WHMIS — Labelling Requirements," http://www.ccohs.ca/oshanswers/legisl/msds_lab.html.

- a reference to appropriate precautionary measures (for example, "store between 15°C and 40°C" or "use in a well-ventilated area"); and

- a description of appropriate first aid measures (for example, "in case of skin contact, flush with water for ten minutes").

In some cases, suppliers or importers include additional information that is not specifically required by law—for example, a phone number and/or address at which the supplier can be contacted, or alternative names (generic name, chemical name, and so on) for the product.

Figure 4.3 shows a sample WHMIS label.

The other disclosure document that a supplier or importer must develop or obtain as a condition of sale of a WHMIS-controlled product is an MSDS. Once developed, the MSDS must be "transmitted" to customers intending to use the product in a workplace as a condition of sale. A sample MSDS is included in appendix D of this book.

An MSDS is a detailed summary of the hazard information for the controlled product. It is required, by law, to include the following nine categories of information:

1. **Product Information.** This must include the name of the product; the name, address, and telephone number of the supplier; and the (usual) use of the product (for example, "agricultural soil additive").

2. **Hazardous Ingredients.** The names, concentration, CAS identification number, and toxicity levels of all hazardous ingredients must be supplied. (The CAS identification number refers to the product's classification in a registry maintained by the Chemical Abstracts Service (see www.cas.org).)

3. **Physical Data.** Schedule II of the CPR lists the following categories of information for inclusion here as:

    - physical state (gas, liquid, or solid)
    - odour and appearance
    - odour threshold
    - specific gravity
    - vapour pressure
    - vapour density
    - evaporation rate
    - boiling point
    - freezing point
    - pH
    - coefficient of water/oil distribution

4. **Fire or Explosion Hazard.** Information about the conditions under which the product may catch fire or explode should be included here, along with recommendations about methods for controlling or containing a fire or explosion.

Figure 4.3 Sample WHMIS Label for Acetone

Callouts (left): Product Identifier; Reference to MSDS; Risk phrases; Precautionary statements; First aid measures
Callouts (top/right/bottom): Hazard symbols; French version; WHMIS hatched border; Supplier identification

ACETONE / ACÉTONE

SEE MATERIAL SAFETY DATA SHEET FOR THIS PRODUCT
VOIR LA FICHE SIGNALÉTIQUE POUR CE PRODUIT

DANGER! EXTREMELY FLAMMABLE. IRRITATES EYES.
PRECAUTIONS: Keep away from heat, sparks, and flames. Ground containers when pouring. Avoid breathing vapours or mists. Avoid eye contact. Avoid prolonged or repeated contact with skin. Wear splash-proof safety goggles or faceshield and butyl rubber gloves. If acetone is present in concentrations greater than 250 ppm, wear a NIOSH-approved respirator with an organic vapour cartridge. Use with adequate ventilation, especially in enclosed areas. Store in a cool, well-ventilated area, away from incompatibles.
FIRST AID: In case of contact with eyes, immediately flush eyes with lots of running water for 15 minutes, lifting the upper and lower eyelids occasionally. Get medical attention immediately. In case of contact with skin, immediately wash skin with lots of soap and water. Remove contaminated clothing and shoes. Get medical attention if irritation persists after washing. Wash clothing before reuse. If inhaled, remove subject to fresh air. Give artificial respiration if not breathing. Get medical attention immediately. If swallowed, contact the Poison Control Centre. Get medical attention immediately. Do not give anything by mouth to an unconscious or convulsing person.
ATTENTION! THIS CONTAINER IS HAZARDOUS WHEN EMPTY. ALL LABELLED HAZARD PRECAUTIONS MUST BE OBSERVED.

DANGER! EXTRÈMEMENT INFLAMMABLE. IRRITE LES YEUX.
MESURES DE PRÉVENTION: Tenir à l'écart de la chaleur, des étincelles et des flammes. Relier les récipients à la terre lors du transvasement. Éviter de respirer les vapeurs ou les brumes. Éviter le contact avec les yeux. Éviter le contact prolongé ou répété avec la peau. Porter des lunettes contre les éclaboussures de produit chimique ou une visière de protection, et des gants en caoutchouc butyle. Si l'acétone est présent en concentration de plus de 250 pour un million, porter un respirateur muni d'une cartouche à vapeur organique approuvé par NIOSH. Utiliser avec suffisamment de ventilation surtout dans les endroits clos. Entreposer dans un endroit frais, bien aéré, à l'écart des produits incompatibles.
PREMIERS SOINS: En cas de contact avec les yeux, rincer immédiatement et copieusement avec de l'eau courante pendant 15 minutes en soulevant les paupières inférieures et supérieures de temps en temps. Obtenir des soins médicaux immédiatement. En cas de contact avec la peau, laver immédiatement la region affectée avec beaucoup d'eau et de savon. Retirer les vêtements et les chaussures contaminées. Si l'irritation persiste après le lavage, obtenir des soins médicaux. Laver les vêtements avant de les réutiliser. En cas d'inhalation, transporter la victime à l'air frais. En cas d'arrêt respiratoire, pratiquer la respiration artificielle. Obtenir des soins médicaux immédiatement. En cas d'ingestion, contacter le Centre de Contrôle des Empoisonnements. Obtenir des soins médicaux immédiatement. Ne rien faire avaler à une victime inconsciente ou en convulsions.
ATTENTION! CE RECIPIENT EST DANGEREUX LORSQU'IL EST VIDE. CHAQUE INDICATION DE DANGER SUR LES ÉTIQUETTES DOIVENT ÊTRE OBSERVÉES.

(BIG) BIG Chemical Company / 123 Nitro Avenue, Vapour Town, BC / 123-4567

An example of a supplier label.

Source: WorkSafe BC, "Symbols & Labels: WHMIS Labels & Checklist," http://www2.worksafebc.com/Topics/WHMIS/SymbolsAndLabels.asp?ReportID=24455.

5. **Reactivity Data**. Information about the product's reactivity with other substances (for example, water, or oxygen) goes here, as does information about any "hazardous decomposition products"—for example, a poisonous gas that would be created should the product react.

6. **Toxicological Properties**. According to the Schedule, this section should include:

   - route of entry, including skin contact, skin absorption, eye contact, inhalation, and ingestion
   - effects of acute exposure to product
   - effects of chronic exposure to product
   - exposure limits
   - irritancy of product
   - sensitization to product
   - carcinogenicity
   - reproductive toxicity
   - teratogenicity
   - mutagenicity
   - name of toxicologically synergistic products

7. **Preventive Measures**. This section should describe measures that users should take to minimize risks to themselves from using the product; for example, appropriate PPE for protection against the product should be mentioned. The section should also explain appropriate measures for cleaning up spills and disposal of containers.

8. **First Aid Measures**. This section should describe any immediate measures that might be useful to minimize harm from a particular kind of exposure (for example, skin contact or inhalation). It should also give information about whether or not immediate medical attention should be sought, and under what circumstances.

9. **Preparation Information**. Finally, the MSDS should include the name and contact information of the party who prepared it, and the date of preparation.

An MSDS is valid for three years from the date of preparation. When it expires, the supplier is required to review the information, make any necessary corrections, and re-date the MSDS. However, if relevant hazard information comes to the attention of the supplier at *any* time—for example, if a product not previously thought to be a teratogen is linked, by a new study, to a birth defect—the supplier is required to prepare a new version of the MSDS immediately for distribution with any subsequent shipments of the product. A form designed to guide the preparation of an MSDS is included in appendix E of this book.

It is important to note that suppliers need not always develop MSDSs "from scratch." Where a product is sold under a trade name but there is an MSDS available for a generic version of the product, the supplier may generally use that version (provided the products are identical). As well, there are many sources of published MSDSs, including the CCOHS. Suppliers who obtain an MSDS from a database like the one maintained by the CCOHS should be aware that the responsibility of ensuring that the MSDS is accurate and up to date rests with the supplier; the supplier must also be sure to take into account all relevant information available about the product. In other words, a supplier cannot simply rely on an MSDS obtained from an outside source without evaluating the accuracy of the content.

AT ISSUE: THE WHMIS AND CONFIDENTIAL BUSINESS INFORMATION

During the consultation process that preceded the introduction of the WHMIS, some stakeholders raised concerns related to the protection of confidential business information.

As you might imagine, some suppliers (and some employers in the field of manufacturing) consider the composition of certain hazardous products of their own creation to be a trade secret. If these manufacturers were required to fully disclose the specific ingredients included in their products, they could potentially lose some business advantage.

To address this concern, the creators of the WHMIS introduced the *Hazardous Materials Information Review Act* (HMIRA). The HMIRA allows a supplier to apply for an exemption or partial exemption from the application of the WHMIS. Where the exemption is required to preserve confidential business information, the supplier will be permitted to withhold certain information from labels or MSDSs.

In general, a supplier will be allowed to withhold specific information about ingredients but will still be required to provide information about the product's properties, the hazards associated with the product, and information about how to minimize exposure risks. For example, a supplier may need to disclose that a product is toxic, the way in which the toxin enters the body (for example, whether it needs to be ingested or whether it is toxic upon inhalation), and the specific toxicity level, but may not need to specify the precise nature of the toxic compound.

Applications for exemptions are heard by the Hazardous Materials Information Review Commission (HMIRC), an administrative tribunal created under the HMIRA.

A similar right to apply for an exemption exists for employers, who may request one under section 40 of the OHSA:

> (1) An employer may file a claim with the claims board for an exemption from disclosing,
>> (a) information required under this Part in a label or material safety data sheet; or
>> (b) the name of a toxicological study used by the employer to prepare a material safety data sheet,
>
> on the grounds that it is confidential business information.

The WHMIS regulation provides, in section 19, that the HMIRC is designated as the "claims board" for the purpose of this section, which means that the HMIRC hears both supplier and employer exemption claims.

Section 40.1 of the OHSA provides that information filed before the HMIRC for the purpose of an application for an exemption is protected by privilege, and cannot be disclosed except in certain narrow circumstances—for example, to a physician in order to assist in a medical diagnosis in an emergency.

WHMIS COMPLIANCE FOR EMPLOYERS

Introduction

As explained above, employer responsibilities under the WHMIS are imposed not by the HPA and CPR, but by the WHMIS provisions in the OHSA and the WHMIS regulation made under the OHSA.

To minimize duplication, government enforcement of both the federal and provincial aspects of the WHMIS, in Ontario, falls to the provincial MOL. The Ministry will include an assessment of the employer's compliance with the WHMIS in its overall health and safety inspections. It also has the power to make orders requiring a non-compliant employer to take steps to comply or requiring that the use of a particular controlled product in the workplace be stopped (OHSA section 33). The sanctions (fines and prison terms) that exist for other kinds of violations under the OHSA are equally available for WHMIS violations. These sanctions are discussed in chapter 10.

FACTS AND FIGURES
MSDS Compliance Statistics

The HMIRC estimates, on its website, that there are over 750,000 product MSDSs in use in Canada at any given moment. In the year 2004–2005, the HMIRC uncovered 2,103 violations of MSDS requirements. The most common error on MSDSs, in that year, was "preparer's name, phone number, date of preparation—at least one missing."

While the labelling requirements placed on employers are more general than those placed on suppliers, employers have significant additional WHMIS obligations in that they are required to ensure that their WHMIS procedures are understood by employees—which means that the procedures are supported by thorough employee training. The heart of the WHMIS is communication: suppliers communicate risks to employers via labelling and MSDSs, and employers communicate risks to employees by making supplier hazard information readily available and by conducting training to support comprehension of that information.

Identification/Classification of Controlled Products

Like suppliers, employers have a duty to determine whether the products they use in the workplace are hazardous, and the specific nature of the hazards (OHSA section 39, WHMIS regulation, section 3). Once a product has been determined to be hazardous, the health and safety representative or JHSC must be informed of the determination.

Although the WHMIS regulation states that employers must use the same procedure for classifying products as is imposed on suppliers under part IV of the CPR (discussed earlier in this chapter), in practice, for employers, these determinations are usually fairly easy to make because employers receive (or should receive) products with appropriate labels on them and accompanying MSDSs.

However, where this is *not* the case, a supplier's non-compliance with the WHMIS does not excuse an employer's. If a product arrives in the workplace without a label or MSDS, it is the employer's responsibility to ask the supplier for these things (OHSA section 41(2)) and to wait to receive them before employees are allowed to use the product. The WHMIS regulation allows employers to store products that do not have labels or MSDSs while they wait (WHMIS regulation, section 5(1)).

Also, where an employer is the producer of a product used in the workplace (for example, a mixture), the employer must, when it first develops the product, give notice of its intention to use the product in the workplace to the MOL (OHSA section 34). The employer must also determine the specific risks associated with the new product so that it can comply with provincial labelling and MSDS requirements (discussed below).

Labels and MSDSs

SUPPLIER-PREPARED DOCUMENTATION

When an employer uses a WHMIS-controlled product (equivalent, under the WHMIS regulation, to a "hazardous material") in a workplace, the employer is obliged to keep the supplier label on the container until the container is empty. If the label becomes illegible or falls off, the employer must replace it, either with a new supplier label or with an employer label (discussed below).

The employer is also required to keep the MSDS (and copies of it for those who want a copy) for the hazardous material in a location that is accessible to the workers and to the Health and Safety Representative or JHSC.

The employer is responsible for keeping an MSDS up to date. This means two things:

1. If three years have passed since the preparation date on the MSDS, the employer is required to obtain an up-to-date version. There is no obligation, for suppliers, to send updated MSDSs to customers; instead, the employer must request the document. If an up-to-date MSDS is not forthcoming from the supplier (for example, where the supplier has gone out of business or has stopped making the product), the employer must obtain an MSDS from someplace else, such as the CCOHS, which maintains a database of MSDSs. If the employer obtains an MSDS from another source or develops one on its own, the employer is responsible for ensuring that the content of the document is accurate.

2. If an existing MSDS has not expired but new hazard information is available (for example, because a new study of the product has been published), the employer is responsible for obtaining a current MSDS from the supplier or for updating the MSDS itself. This must to be done within 90 days of the new information becoming available.

EMPLOYER LABELS

When an employer decants a hazardous product into its own container (for example, a smaller bottle for ease of use) or where a supplier label on an original container becomes illegible or falls off, the employer can apply an employer label to the container.

In some narrow circumstances—where an employee fills a portable container (for example, a paint tray) from a labelled container and uses the portable container exclusively (without sharing) during a single shift so that it is used up or emptied at the end of the shift—the employer is exempt from labelling the container (WHMIS regulation, section 10).

The obligation to label hazardous materials extends both to containers (according to the WHMIS regulation, section 1, "a bag, barrel, bottle, box, can, cylinder, drum, storage tank or similar package or receptacle") and to other material containment systems (for example, pipes and piping systems, tanks and tanker vehicles, conveyor belts, and process or reaction vessels, according to the WHMIS regulation, section 11). In the case of these systems, an employer can use colour coding or placards as an alternative to labels to identify contents.

The information included on a workplace label is less closely regulated than that of a supplier label. According to WHMIS regulation, section 1 (definition of "workplace label"), a label must disclose the following:

> (a) a product identifier identical to that found on the material safety data sheet for the controlled product,
> (b) information for the safe handling of the controlled product, and
> (c) that a material safety data sheet, if supplied or produced, is available.

Of course, employers are always welcome to provide more detailed information, if they so choose, about the hazardous material on their labels—and many do. Additional details can be useful for workers and can reinforce the content of employee training, which is discussed in the next section.

Worker Training

Section 42 of the OHSA requires that, where workers will be exposed to hazardous materials or agents in their work, the employer provide "instruction or training" about the material or agent and about how to minimize exposure risks.

This training is to be developed in consultation with the Health and Safety Representative or JHSC for the workplace. The training, or the employee's understanding of the learning principles, must be reviewed at least annually. However, if there is a "change in circumstances" (for example, a change in work processes that involves the use of new hazardous materials or the use of existing materials in different ways), retraining should be done sooner.

The specific requirements of a hazardous materials training program are described in the WHMIS regulation. Section 6 provides that employees likely to be exposed should be made aware not only of all the information provided by the supplier of the product (in other words, the content of the label and MSDS), but also of all additional information of which the employer ought reasonably to be aware (usually, published studies). The employer, for example, may use a product for a purpose other than that anticipated by the supplier, and may have learned of risks associated with that use that are not mentioned on the label or in the MSDS.

Section 7 of the WHMIS regulation lists the components of a complete education program as follows:

7.(1) An employer shall ensure that every worker who works with or in proximity to a controlled product is instructed in,

(a) the contents required on a supplier label and workplace label, and the purpose and significance of the information contained on the labels;

(b) the contents required on a material safety data sheet and the purpose and significance of the information contained on a material safety data sheet;

(c) procedures for the safe use, storage, handling and disposal of a controlled product;

(d) procedures for the safe use, storage, handling and disposal of a controlled product when it is contained or transferred in,

(i) a pipe,
(ii) a piping system including valves,
(iii) a process vessel,
(iv) a reaction vessel, or
(v) a tank car, a tank truck, an ore car, a conveyor belt or a similar conveyance;

(e) procedures to be followed when fugitive emissions [leaks, including vapour or fume leaks] are present; and

(f) procedures to be followed in case of an emergency involving a controlled product.

The training provided to employees must be specific to their particular workplace and procedures. For example, employees who work with solvents in a particular room might be instructed to "open the windows on the east and west walls while using the solvent and leave them open for an hour afterward." Employees should also be informed of the location, within the workplace, of safety gear and supplies, and contact information for first-aiders.

Finally, employers must ensure not only that training has been delivered, but also that it is likely to accomplish the safety goals for which it has been designed. This is usually accomplished through oral or written testing. If employees fare poorly on tests, efforts should be made to change the training so that it is more effective.

KEY TERMS

Material Safety Data Sheet (MSDS)

REVIEW QUESTIONS

1. Many commentators describe the WHMIS as primarily a communications program. Do you agree with this? Why or why not?

2. Why did the creators of the WHMIS feel it was important to prescribe classification information under federal rather than provincial law?

3. As a supplier of hazardous products, what steps do you think you would need to take to comply with the requirement to include, in the MSDSs you prepare, "all information about the product of which the supplier is aware or ought reasonably to be aware"?

4. How does the WHMIS seek to balance the goal of worker safety with the need to protect confidential business information?

5. How, as an employer, might you design the delivery of WHMIS training so that new employees receive training immediately while existing employees are given opportunities to refresh their knowledge?

6. As an employer, how might you make decisions about workplace label content and design? For example, is more information always better, or are simpler labels more likely to be read?

CHAPTER 5

Compensation for Injured Workers

CHAPTER OBJECTIVES

After completing this chapter, you should be able to:

- Understand which legislation governs compensation for injured workers and which government agency administers the insurance fund that pays this compensation.
- Explain which employers must insure their workers through the *Workplace Safety and Insurance Act* fund, which employers must pay worker claims directly, and which employers can choose to insure their workers (voluntary participation).
- Describe how employer premiums are calculated under the legislation.
- Understand how employers can manage their premiums through good safety performance.
- List the benefits available to injured and sickened workers.
- Explain how fraud and non-compliance are dealt with by the WSIB.

INTRODUCTION

As you learned in chapter 2, the earliest examples of workplace safety legislation were aimed at compensating injured workers for their injuries and providing survivor benefits to spouses of workers killed on the job. Ontario's first worker compensation statute, modelled after British law, was adopted in 1886. Under that legislation, compensation was funded solely by employers, and paid modest benefits only after a finding of employer fault.

Workers' compensation has come a long way since 1886. Ontario's current legislation, the ***Workplace Safety and Insurance Act* (WSIA)**, enacted in 1997, creates a sophisticated no-fault insurance scheme, funded by employers on behalf of workers. The scheme includes features designed to motivate good safety-promotion performance on the part of employers, and to support workers in their quest to get back to work.

***Workplace Safety and Insurance Act* (WSIA)**
the statute that governs the application of the Ontario government's scheme for the compensation of workers injured or sickened on the job

This chapter provides an overview of the WSIA, with an emphasis on the employer's perspective on issues such as how to minimize insurance costs. Managing injured workers' return to work is covered in detail in chapter 11.

> ### FACTS AND FIGURES
> #### Injuries and Occupational Illnesses Before 1998
> The current workplace health and safety legislation, the *Workplace Safety and Insurance Act* (WSIA), came into force on January 1, 1998. Before that time, workers' compensation was governed by the *Workers' Compensation Act* (WCA). There are significant differences between the two legislative schemes, and it is important to know that workers whose injuries occurred (or who contracted an occupational illness) prior to January 1, 1998 and who made their claims for benefits prior to that date are still governed by the WCA, and not by the current legislation.

OVERVIEW OF THE LEGISLATION

Application

GENERAL

The WSIA applies to many workers and workplaces in Ontario. According to section 1 of the Act, its four express purposes are:

1. To promote health and safety in workplaces and to prevent and reduce the occurrence of workplace injuries and occupational diseases.

2. To facilitate the return to work and recovery of workers who sustain personal injury arising out of and in the course of employment or who suffer from an occupational disease.

3. To facilitate the re-entry into the labour market of workers and spouses of deceased workers.

4. To provide compensation and other benefits to workers and to the survivors of deceased workers.

Within ten days of hiring its first worker or acquiring a business that has workers, an employer must register with the Workplace Safety and Insurance Board (WSIB).

THE SCHEDULES

An employer's duties, beyond registration, are largely dependent upon which Schedule the employer falls into. There are two schedules (essentially, lists) of employers under the legislation. These can be found in the General regulation (O. Reg. 175/98) made under the Act.

Schedule 1 includes those employers who are liable to pay **premiums** to fund the WSIA insurance scheme. Employers in this category are those whose business involves:

- **Class A—Forest Products**. This class includes lumbering; manufacturing certain kinds of wooden paper and cardboard materials (such as plywood);

premium
an amount of money paid for an insurance policy, usually at regular intervals

operating lath mills, sawmills, shingle mills, pulp mills, paper mills and associated yards; and kiln-drying and creosoting of timber.

- **Class B—Mining and Related Industries**. This class includes, for example, coal and diamond mining; marble quarrying; operating gravel and sand pits (and associated crushing, and so on); manufacturing peat fuel; and stone cutting.

- **Class C—Other Primary Industries**. These include, for example, most kinds of agriculture and farming; fishing; landscaping; and well drilling.

- **Class D—Manufacturing**. This class includes manufacturers of almost every consumer item imaginable, excluding things that fall into classes A through C.

- **Class E—Transportation and Storage**. This class includes, for example, trucking, hauling, taxi companies, railways not included in Schedule 2, air transport companies, docks and wharves; and, on the storage side of things, running grain elevators, warehouses, and other storage sites and equipment.

- **Class F—Retail and Wholesale Trades**. This class includes a very wide range of sales and service-based businesses of various kinds, such as new and used automobile sales, machinery sales and service, gas stations, most kinds of distributors that do not manufacture the goods distributed, and the operation of a retail mercantile business (a store of almost any kind).

- **Class G—Construction**. This class includes all kinds of construction, as well as activities that support construction, such as demolition; installation of boilers, engines, and so on; and trades such as electrical wiring, gas welding, and plumbing.

- **Class H—Government and Related Services**. This class includes the installation of infrastructure, such as power lines, pipelines, telephone lines, and cable; the delivery of natural gas and electricity; and the operation of hospitals and medical laboratories.

- **Class I—Other Services**. This class includes a number of service businesses, such as dry cleaners and laundries, "photographic finishing", operating a telephone service, hotels and motels, operating office buildings, advertising-agency work, the work of mining consultants and engineers of many kinds, restaurants and catering, non-medical laboratories, and private security services.

It is important to note that the determination of which category a business falls into is quite complicated, because the wording of the Schedule is general enough to suggest overlap between classes. For example, while maintenance of telephone lines "as a business" falls into Class H, "operation of a telephone service" falls into Class I. Similarly, operating a business that sells and services machinery falls into Class F, while *renting* machinery falls into Class I—except for boats, the rental of which falls into Class F.

Employers who need to know how their Schedule 1 operations are classified should read the Schedule carefully, but it is probably wise to contact the WSIB for confirmation of the specific classification. Knowing one's classification is important, because the way in which a business or industry is classified determines the premiums charged. Premiums differ from class to class to reflect the different levels of risk associated with different industries.

Schedule 2 employers are liable to pay the *entire* cost of any benefits payable, under the WSIA, to workers whom they employed at the time of the workers' injuries. These employers are assessed payments by the WSIB and pay the necessary funds to the WSIB, reimbursing it for payments made to workers.

Schedule 2 employers are connected with government or public services. They include, for example:

- railways;
- telephone and telegraph service infrastructure;
- municipal corporations;
- public library boards;
- shipping companies;
- the constructor-owners (that is, not private contractors) of interprovincial and international bridges; and
- international airlines.

For an exhaustive list, see the General regulation (O. Reg. 175/98).

FACTS AND FIGURES

Excluded Industries

There are certain industries specifically *excluded* from schedules 1 and 2. These are listed in the General regulation made under the WSIA as follows:

1. Barbering and shoeshining establishments.
2. Educational work, veterinary work, and dentistry.
3. Funeral directing and embalming.
4. The business of a photographer.
5. Taxidermy.

The schedules themselves are complicated enough, but the task of determining how a business will be classified by the WSIB does not end there. The General regulation sets out rules to govern situations that may be confusing to employers. These can include cases:

- where an employer conducts a Schedule 1 activity (for example, construction) in support of its main business, which is more like a Schedule 2 business (in this case, the employer will be deemed to fall into Schedule 1);
- where an activity that falls into a Schedule is carried out in the context of bankruptcy or insolvency (under the direction of a trustee in bankruptcy, for example); or
- where an employer conducts an activity only partly as a business—consider, for example, a hobby farmer who grows vegetables for his family (while working as an employee of a phone company, say) but who hires farm hands to help with the harvest, a portion of which he sells at a farmer's market.

If an employer has any confusion whatsoever about how the schedules apply to its business, it is advisable that the employer read the General regulation in detail as well as contact the WSIB for confirmation of the business's classification.

If the employer disagrees with its classification under the system of schedules, the employer can appeal the WSIB's decision before the Workplace Safety and Insurance Appeals Tribunal (WSIAT).

Employers who do *not* fall into the classification scheme of either Schedule are not required to obtain insurance for their workers by paying premiums. People who work for these employers are not eligible for benefits under the plan.

Who Is a "Worker"?

In recognition of the fact that the risk of injury is unrelated to the details of the working arrangement, the WSIA provides a very wide definition of "worker." Under the legislation, a worker can be:

a person who has entered into or is employed under a contract of service or apprenticeship and includes the following:

1. A learner.
2. A student.
3. An auxiliary member of a police force.
4. A member of a volunteer ambulance brigade.
5. A member of a municipal volunteer fire brigade whose membership has been approved by the chief of the fire department or by a person authorized to do so by the entity responsible for the brigade.
6. A person summoned to assist in controlling or extinguishing a fire by an authority empowered to do so.
7. A person who assists in a search and rescue operation at the request of and under the direction of a member of the Ontario Provincial Police.
8. A person who assists in connection with an emergency that has been declared by the Lieutenant Governor in Council or the Premier under section 7.0.1 of the *Emergency Management and Civil Protection Act* or by the head of council of a municipality under section 4 of that Act.
9. A person deemed to be a worker of an employer by a direction or order of the Board.
10. A person deemed to be a worker under section 12.
11. A pupil deemed to be a worker under the *Education Act*.

As you can see, almost anyone who performs a service on behalf of or in assistance to someone else is a "worker" for the purpose of the WSIA.

How the Fund Works

Employers pay annual premiums assessed by the WSIB. These premiums are invested by the WSIB to create an insurance fund.

"COMPENSABLE INJURIES"

When an employee sustains a "compensable injury" as defined by the legislation, she must file a claim for benefits with the WSIB. A compensable injury is defined in the legislation as a "personal injury by accident arising out of and in the course of his or her employment."

Whether or not an accident is "in the course of" a worker's employment depends on where the worker was when it happened, when it happened (during working hours or not), and on the activity the worker was doing when it happened.

Whether or not an accident "arises out of" employment depends on whether the activity the worker was pursuing at the time was related to his assigned work duties.

HARD LESSONS
In the Line of Duty?

It is important to understand that the definition of a compensable injury "arising out of" employment is broad enough to include medical, emotional, and psychiatric problems that may arise much later but that are directly related to an injury sustained at work.

For example, in a case (*Decision No. 1726/06* (2006)), a 26-year-old man sustained a serious neck injury at work in 1976. The injury aggravated a congenital condition that led, eventually, to brain hemorrhage, repeated surgeries, surgical complications, and chronic pain. By the early 2000s—about 25 years after the original injury—the patient was having to take psychotropic medications for post-traumatic stress disorder. The Workplace Safety and Insurance Appeals Tribunal (WSIAT) allowed the worker's claim for the psychiatric disability.

In another case, a woman suffered a compensable back injury in 1990 while lifting a tray. After suffering another injury in 1991 she began to develop symptoms of depression and anxiety, and stopped working permanently in 1992. In 2000 she applied for benefits based on psychotraumatic disability. Even though the WSIAT found that the pain that led to the psychiatric problems was only partly related to the back injury (she also had pain from pre-existing neck and shoulder problems), because the compensable injury contributed to the cause of the psychotraumatic disability she was eligible for benefits. The WSIAT made this decision despite the fact that there is a general guideline in place, in cases of psychiatric disability, to restrict recovery to cases where the psychiatric problems arise within five years of the injury (in this case, the worker was not diagnosed with major depression until ten years after her injury). The WSIAT wrote:

> We find that the totality of the evidence in this case, as discussed below, makes it inappropriate to apply the 5-year guideline. ... We find that the worker's depression is a result of her poor physical health and the impact of her poor health on the other areas of her life.

Source: *Decision No. 954/06* (2006).

For the purpose of WSIB benefits, there is an important difference between psychiatric disorders that arise out of physical injuries, and mental stress that is not associated with an injury. (See the discussion of the WSIB's policy with respect to mental stress in the Facts and Figures feature later in this chapter.)

In some cases, the way in which a worker's injury affects his or her overall mental health can be related to feelings about his work life. If pain prevents the worker from doing work activities and the employer is not very supportive in supporting a successful return to work, the worker may quit. Not being able to work can make many people feel helpless, overdependent on others, and disconnected from the means to make a valuable contribution to society. These feelings can exacerbate psychiatric symptoms.

By making an effort to fully support returning workers so that they can find something productive to do that does not cause excessive pain, employers may be able to help minimize the incidence of claims for psychiatric disability—and fewer claims means lower premiums.

The WSIA also defines what is meant by an "accident." An accident is:

- a chance event occasioned by a physical or natural cause (this is interpreted broadly to include many kinds of "events", for example, machinery failure, fires, spills, leaks, and so on);

- a willful and intentional act that is *not* the act of the worker (for example, workplace violence, or another employee's decision to intentionally "cut corners" and ignore a safety rule); or

- disablement arising out of and in the course of employment (this has been interpreted to include repetitive-strain related injuries, back pain, and other injuries that do not have an acute onset—that is, where the precise time of the "accident" cannot be pinpointed).

It is important for employers to remember that before the advent of worker compensation funds, many of these kinds of "accidents" might have formed the basis of lawsuits brought by workers against their employers. By requiring workers to give up their rights to sue for damages due to injury, the WSIA makes the cost of compensating injured workers a predictable and budgetable—rather than a capricious—expense for businesses.

COMPENSATION FOR OCCUPATIONAL DISEASES

Under the *Workers' Compensation Act* (the legislation that came before the WSIA), workers were entitled to compensation for specific "industrial diseases" recognized in the legislation.

The WSIA has replaced that fairly narrow approach by allowing workers to be compensated for occupational diseases that meet the following general definition:

"occupational disease" includes,
(a) a disease resulting from exposure to a substance relating to a particular process, trade, or occupation in an industry,
(b) a disease peculiar to or characteristic of a particular industrial process, trade, or occupation,
(c) a medical condition that in the opinion of the Board requires a worker to be removed either temporarily or permanently from exposure to a substance because the condition may be a precursor to an occupational disease, or
(d) a disease mentioned in Schedule 3 or 4.

Schedules 3 and 4 create legal presumptions that when a worker has a disease that is associated with a specific work process and the worker has been engaged in that work process, that the work process caused the disease. For example, if the worker has asbestosis and her job involves demolition work that causes asbestos particles to be made airborne, then the WSIB is entitled to presume that the worker's work caused the asbestosis (and that the asbestosis qualifies as an occupational disease). Schedules 3 and 4 (combined) create 34 different process/disease presumptions.

The WSIA compensates workers for occupational disease-related disability in a similar way to how it compensates for injuries. Section 15(2) explains that a worker suffering from (and impaired by) an occupational disease will be compensated as if the disease were an injury caused by an accident. For the purpose of determining the details of the claim, the beginning of the worker's period of impairment is considered to be the time of the accident.

FACTS AND FIGURES

Limited Compensation for Mental Stress

Mental stress is very difficult to diagnose, quantify, and attribute to a specific cause. People differ widely in their susceptibility to stress and their ability to recover from it. For this reason, the WSIA limits the availability of benefits for mental stress to fairly narrow circumstances, as described in section 11(5):

> A worker is entitled to benefits for mental stress that is an acute reaction to a sudden and unexpected traumatic event arising out of and in the course of his or her employment. However, the worker is not entitled to benefits for mental stress caused by his or her employer's decisions or actions relating to the worker's employment, including a decision to change the work to be performed or the working conditions, to discipline the worker or to terminate the employment.

The employer's role in assisting with employee claims for compensation is discussed in the next section. The benefits available to a worker upon having her claim accepted are discussed later in this chapter.

Employer Duties Under the WSIA

Employers have a number of duties under the WSIA. These include:

- to keep a safe and well-maintained workplace;
- to provide hazard information, proper safety equipment, training, and competent supervision;
- to provide first aid training, services, and supplies;
- to post the WSIB's Form 82 poster, "In Case of Injury at Work," in a prominent place (see figure 5.1);
- to post a copy of the WSIA where workers can consult it;
- to pay WSIA premiums, if applicable;
- to provide the WSIB with the information necessary for the calculation of premiums (for example, payroll information); and
- to report workplace illnesses and injuries to the WSIB.

Figure 5.1 "In Case of Injury at Work" WSIB Form 82 Poster

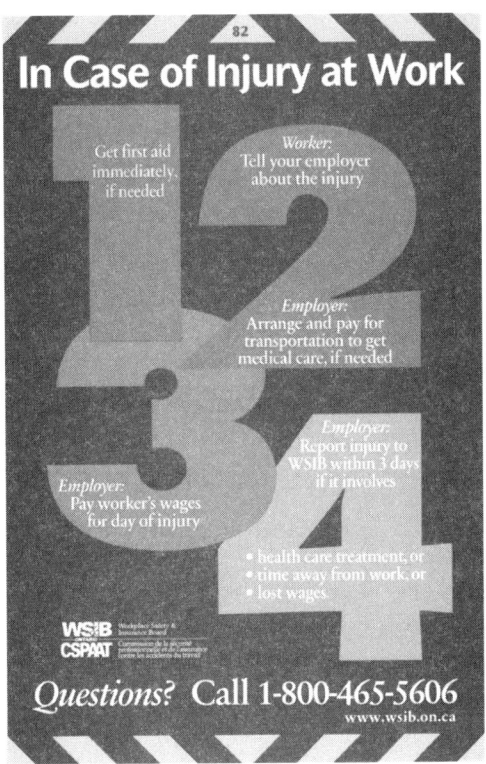

Source: Workplace Safety and Insurance Board, "Prevention—In Case of Injury at Work Poster (Form 82)," http://www.wsib.on.ca/wsib/wsibsite.nsf/Public/InCaseOfInjuryPoster.

Once an employer has been registered and classified by the WSIB, it can, if it so chooses, sign up for a "Calculate and Report Premium 'eService.'" The service allows employers to calculate premiums, report payroll, and pay premiums electronically.

Workers also have duties under other legislation that help support their claims under the WSIA. For example, where workers are required to work with materials that are "designated substances" under the WHMIS, employers must keep records of spills and other exposures that have the potential to lead to injury or illness for workers.

Understanding and Managing Premiums

The WSIB explains, on its website, that the WSIA premium payable by an employer depends on "the health and safety risk of your type of business, the size of your payroll, and on your company's health and safety record."

HOW PREMIUMS ARE CALCULATED

The WSIB groups industries that are similar from an injury/illness risk perspective into classification units (CUs). Each CU is assigned to a rate group. Basic premiums are assessed annually for each rate group, with groups with a higher risk of injury being assigned higher base premiums.

An individual employer's premium is based on the general rate group assessment. A discount is sometimes applied for better-than-average safety performance, or a penalty (in the form of an individual premium higher than the base amount) for employers with a poor safety performance.

FACTS AND FIGURES

Factors Influencing Premiums

Section 82 of the WSIA has the following to say about the grounds on which a particular business might be assessed a premium above or below the base premium for its rate group:

> The Board may increase or decrease the premiums otherwise payable by a particular employer in such circumstances as the Board considers appropriate including the following:
>
> 1. If, in the opinion of the Board, the employer has not taken sufficient precautions to prevent accidents to workers or the working conditions are not safe for workers.
>
> 2. If the employer's accident record has been consistently good and the employer's ways, works, machinery and appliances conform to modern standards so as to reduce the hazard of accidents to a minimum.
>
> 3. If the employer has complied with the regulations made under this Act or the *Occupational Health and Safety Act* respecting first aid.
>
> 4. If the frequency of work injuries among the employer's workers and the accident cost of those injuries is consistently higher than that of the average in the industry in which the employer is engaged.

For example, a car wash operation falls into the CU coded 6391-000, which is assigned to rate group 630, "Vehicle Services and Repairs." The basic 2007 premium for this rate group was $3.40 per $100 of payroll.

It is possible for a business to fall into more than one rate group if it carries on more than one kind of business activity. For example, a car wash may be offered by a business that also operates a gas bar. Gas bars belong to CU 6331-001, which falls into rate group 633, "Petroleum Products, Sales." For this group, the basic 2007 premium was about $2.12 per $100 of payroll.

Where a business has operations that fall into two different rate groups, the business can report earnings (and pay premiums) under each one separately. To do this, the business must keep accurate payroll records. The WSIB will be alert to any attempt to save money by underreporting in the higher-premium rate group and overreporting in the lower-premium rate group.

Premiums have three components:

1. costs of new injuries or illnesses;
2. a charge for overhead costs; and
3. an unfunded liability amortization charge.

The bulk of the premium falls under the first component. It is based on a projection, based on past experience, of the actual future cost of any injuries and illnesses that can be expected to arise in the premium year. The second component is designed to offset the administrative costs of running the fund and paying claims. The third component, the "unfunded liability amortization charge," is a charge designed to remedy the current situation under which there is not enough money currently in the fund to pay the projected future cost of existing claims. As of 2005, when the most recent funding framework was designed, the unfunded liabilities facing the fund were reported as about $6.4 billion. The current funding framework is expected to result in the fund being fully funded (with the unfunded liability eliminated) by 2014.

Employers interested in the specifics of the Workplace Safety and Insurance Fund's funding framework should consult the *Funding Framework* document available on the WSIB's website.

The easiest way for an employer to determine its classification for the purpose of WSIA premiums is to contact the WSIB directly. The WSIB maintains two key contact lists on its website:

- Employers with small businesses (under 20 employees) should consult the Office Locator.
- Larger employers should consult the list of Industry Sector contacts.

CHALLENGING YOUR CLASSIFICATION

Employers can work to manage their WSIA premiums in two key ways:

1. by ensuring that their operations are properly classified under the WSIB system, and
2. by optimizing safety performance (and documenting good performance) in an attempt to win an incentive-based adjustment.

Employers who disagree with the classification assigned to them by the WSIB should consult the Employer Classification Manual, available on the WSIB's website or in print form from the WSIB. Once the employer has identified an alternative classification that it believes is more appropriate for its operation(s), the employer should contact the WSIB informally and discuss the matter. If the employer is not satisfied with the WSIB's response, he or she can make a formal request for a premium adjustment. According to the WSIB's *Operational Policy Manual*, such a request must be made in writing and the employer must:

- clearly identify the premium issue requiring an adjustment, and
- specify their reasons for requesting the adjustment, e.g. why they believe there was a mistake made in the calculation of their premium, their classification, or other charges applied to their account, etc.

The WSIB will then make a decision about whether or not to adjust the premium. Once the WSIB has made a decision about whether or not to reclassify the business, it will send a letter to the employer explaining its decision. The letter will include an invitation to the employer to contact the WSIB for an informal discussion of the decision.

If the informal discussion does not satisfy the employer, he or she has the right to fill out an objection form and file it with the WSIB. This form will then be the basis of a formal reconsideration of the decision. Conflicts that arise between employers and the WSIB are generally resolved with the help of an Appeals Resolution Officer (ARO) at the WSIB.

The WSIB has the power, under section 121 of the WSIA, to reconsider its own decisions (for example, a decision with respect to a business's classification).

> ## FACTS AND FIGURES
>
> ### Appeals Resolution Officers at the WSIB
>
> In 1998 the WSIB introduced changes to the way it handles reconsiderations and appeals. Prior to 1998 three different kinds of WSIB administrators—Mediators, Reinstatement Officers, and Appeals Officers—were involved in the appeals process. The functions of these three administrators have now been merged into the single role of the Appeals Resolution Officer, or ARO.

If the ARO declines to reconsider the employer's request and the employer is not satisfied with the WSIB's explanation, the file will be referred to the Appeals Branch. There are a range of resolution procedures available at the Appeals Branch, and these will be explained to the business owner (or his or her lawyer) by the ARO. They are also described in the WSIB's *Appeal System Practice & Procedures* manual, which is available on the WSIB's website.

If the decision arrived at in the Appeals Branch does not satisfy the employer, the employer can bring a further appeal before the Workplace Safety and Insurance Appeals Tribunal (WSIAT). The WSIAT is an administrative tribunal, not a court, but it conducts fairly formal **hearings**—often with witness testimony—and issues a decision much in the same way as a court would issue a judgment.

hearing
the listening to evidence and pleadings in a court or other officially constituted body

Business owners who are bringing an appeal before the WSIAT should hire legal counsel to handle the appeal (if they have not already been working with counsel for the internal appeal). In preparing for the appeal, HR staff who may be involved in the case may want to review the WSIAT's procedures (available on the WSIAT's website at www.wsiat.on.ca).

> ## PROFILE
>
> ### Office of the Employer Adviser
>
> The Office of the Employer Adviser (OEA) is a government agency dedicated to assisting employers with workplace safety and insurance issues. It is funded by premiums and administrative charges collected by the WSIB, and so its services are offered free of charge to employers.
>
> The OEA is intended as a resource for employers, providing advice to employers of all sizes and representation to employers with fewer than 100 employees.
>
> The advice centre of the OEA is accessible by telephone at 1-800-387-0774 (416-327-0020 in Toronto). OEA staff can also answer questions via email at weboea@mol.gov.on.ca. Publications with information on many aspects of workplace safety management can be accessed via the agency's website at www.employeradviser.ca.
>
> In addition to advice, the OEA offers mediation, negotiation, and representation services. These services are generally available only to smaller employers (those with fewer than 100 employees) who may not be able to afford a private lawyer.

If an employer is required to argue an appeal either within the WSIB appeals system or before the WSIAT, the employer can contact the OEA to seek help with the appeal. OEA staff have considerable experience with resolving a wide range of workplace safety insurance issues and can help employers solve problems quickly so that they—and their employees—can get back to work.

HARD LESSONS
Classification in Special Cases

There is enormous variety in the world of business, and not every employer falls neatly into a WSIB classification. Where a review of the existing classifications reveals no obvious "fit," the employer and the WSIB are forced to revisit the philosophies that underlie the classification system in order to come up with a solution that makes good common sense.

In a particular 2006 WSIAT case, an employer challenged its classification as "Other Services Incidental to Agriculture." The employer, a not-for-profit organization, had no competitors and offered a one-of-a-kind service to dairy farmers. While all milk is subject to compulsory testing before it can be marketed, this testing is done on bulk shipments of milk. The employer in this case offered *voluntary* testing of the milk of individual cows. The testing was designed to allow farmers to identify low producers and cows whose milk was below the overall standards of the whole herd. While the employer did laboratory testing (which usually falls under "Other Services Incidental to Agriculture"), only 8 or 9 employees worked in the lab, with an additional 140 or so working in the head office and as customer service representatives.

In its submissions to the WSIAT, the employer argued that its work had little in common with other businesses in its assigned class. In addition, its safety performance—which it described with reference to the NEER system discussed later in this chapter—was significantly better than the average for its class (and possibly more than ten times better).

While the comparison of safety performance is not recognized by WSIB as a deciding factor in an analysis of a firm's classification, the WSIAT recognized that it had some evidentiary weight. Considering this evidence along with all of the other evidence about the employer's business, the WSIAT ruled that the employer should be reclassified as "Agricultural Management and Consulting Services," a classification that would put it in a rate group that enjoyed lower premiums. (For more details, see WSIAT *Decision No. 1813/06*, issued November 15, 2006.)

MANAGING PREMIUMS THROUGH GOOD SAFETY PERFORMANCE

Section 83 of the WSIA provides that the WSIB "may establish experience and merit rating programs to encourage employers to reduce injuries and occupational diseases and to encourage workers' return to work." The WSIB has, in fact, done this.

One of the rating programs employed by the WSIB is the New Experimental Experience Rating (NEER) system. Employers who are in a non-construction rate group and who pay more than $25,000 per year in premiums are required to participate in NEER.

The NEER program works by comparing a business's safety track record to that of other similar businesses in the same rate group—for example, dry cleaners are compared to other dry cleaners.

If a business's cost of claims in a particular year is higher than the industry average, it may have to pay a premium surcharge; if the cost of claims is lower than the industry average, it may receive a premium rebate. The comparison is adjusted for company size, and the percentage of surcharges and rebates applied to small companies is reduced to reflect random "blips" in injury rates that do not truly reflect safety performance.

For detailed information on how claim costs, rebates, and surcharges are calculated, employers should review the WSIB's guide to NEER, available online or from WSIB offices.

For construction industry businesses with average annual premiums over $25,000, the WSIB offers a similar program called CAD-7; again, more information about this program is available from the WSIB.

✲ Success Story

Benefits for Employers, Too

As an employer, it is important to realize that the insurance scheme created by the WSIA is not just a cost of doing business. The WSIA system provides important benefits to employers, too, including:

- **Protection against expensive litigation.** By paying (predictable) premiums, employers can avoid the financial uncertainty that would result from workers having to sue for compensation in the civil courts.

- **No public fault-finding.** Where workers have access to benefits under the WSIA system, as long as there is no prosecution under the OHSA, there is no public finding of fault against the employer, and no associated negative press.

- **Prevention programs.** The WSIB supports employers in reducing lost-time injuries by offering injury and illness prevention programs for workers and supervisors.

- **Assistance with return to work.** The WSIB offers resources to assist employers in welcoming recovering workers back to work.

- **Optional coverage for management.** The WSIB offers optional coverage for employers themselves, should they wish to insure against the risk of their own injuries or illnesses.

Finally, for small businesses, the applicable WSIB incentive program is the Merit-Adjusted Premium Plan for Small Business (MAP). MAP applies to businesses that pay between $1,000 and $25,000 in WSIA premiums each year. Like the NEER and CAD-7 programs, MAP offers decreased premiums for good health and safety performers and imposes increases on those who perform poorly. Poor performance is measured according to two criteria: the number of claims and claim severity. Claims under $500 do not affect a company's rating for the purpose of increases.

Understanding Benefits

The WSIA offers several different categories of benefits for injured workers. Which benefits a worker will receive depends on the nature of his injury or illness, his medical needs, and his life stage and family status. Most WSIA benefits are paid to the worker himself, but some are paid instead to surviving family members. Figure 5.2 provides a summary of benefits available.

MANAGING WORKERS' RETURN TO WORK

As an employer, you have a duty to support injured workers' "early and safe" return to work. Managing this process is the subject of chapter 11 of this book. For now, the following is a summary of some of the specific obligations that you may have toward injured or ill members of your workforce.

- You should contact the worker as soon as possible after the accident and remain in regular contact with her while she is recovering.
- You should inquire about the worker's health and her functional abilities and obtain a written consent to obtain and/or share any personal details, if necessary, to support her return to work (for example, you may need to prepare a file on the worker and her injury for an in-house occupational therapist).
- You should provide the worker with a Functional Abilities Form (available on the WSIB's website) to take to her health care provider for completion.
- You should maintain an ongoing discussion with the worker about her return to work, including a discussion of estimated date of return.
- In preparation for the worker's return, you should put in place equipment and/or processes that will accommodate any new limits on the worker's functional abilities (for example, if the worker will have difficulty climbing stairs, you might consider moving her office to the ground floor).
- If the worker is expected to have a permanent or long-term disability, you should educate yourself about the type of disability, about the obstacles she is likely to encounter in her work because of the disability, and about potential solutions to these obstacles.

These steps and others will be discussed in greater detail in chapter 11.

MANAGING FRAUD AND SUSPECTED FRAUD IN THE WSIA SYSTEM

Occasionally, the WSIB is faced with **fraud** on the part of a business owner or worker. If the WSIB suspects fraud, there may be an investigation, and human resources personnel may be required to participate by providing information or by undergoing an **audit**.

Worker Fraud and Non-Compliance

The most common fraud perpetrated by workers within the WSIA system is misrepresentation of entitlement to benefits. A worker may attempt to exaggerate his or her injury, may make a claim based on an impairment that does not exist, or may attempt to continue collecting benefits after an injury or illness has resolved.

fraud
the act of obtaining money or some other benefit illegally through deliberate deception

audit
in the context of the WSIA, a compulsory review by Workplace Safety and Insurance Board investigators of an employer's compliance or lack of compliance with the WSIA

Figure 5.2 Workplace Safety and Insurance Benefits Overview

Benefit Name	Paid To	Eligibility Criteria	Benefit Amount and Duration
Loss of Earnings (LOE) Benefits	Worker	Work illness or injury that meets the WSIB definition; employer is insured	For injuries after 01/01/98, 85% of take home pay up to the annual wage ceiling ($71,800 in 2007). For injuries before 01/01/98, rates are different (90% or 75%, depending on injury date). Benefits are paid until the impairment resolves or the worker turns 65.
Non-Economic Loss (NEL) Benefit	Worker	Worker suffers an injury resulting in permanent impairment that causes physical, psychological, or functional loss (affects quality of life)	The (lump sum) benefit is calculated at the point where no further improvement to the worker's condition is expected. At that time, the degree of permanent impairment is estimated in the form of a percentage. The percentage is multiplied by a base amount ($52,156.06 in 2007), and then adjusted for the worker's age (increased for each year under 45, decreased for each year over 45).
Loss of Retirement Income (LRI) Benefit	Worker, or her dependants (if worker dies before age 65)	A worker continues to receive LOE benefits past the 12-month mark	After the first 12 months of LOE benefits, the WSIB sets aside an amount equal to 5% of the worker's LOE benefits. The WSIB holds these benefits for the worker and invests them (workers can contribute money to the fund if they want to). When the worker reaches age 65, the money that was set aside and any accumulated investment interest is paid to the worker or her dependants.
Benefit for Future Economic Loss (FEL)	Worker	Only workers injured between 1990 and 1998 who were permanently impaired	90% of the difference, if any, between the worker's take-home pay before the permanent impairment and the worker's expected take-home pay after the impairment
Health Care Benefits	Worker	All injured workers who need treatment	Most health care expenses, including medical care, prescription drugs, assistive devices, and injury-related transportation costs; most of these are paid for directly by the fund

Benefit Name	Paid To	Eligibility Criteria	Benefit Amount and Duration
Access to the Occupational Disease and Survivor Benefits Program	Worker and his or her survivors	Workers who are afflicted with a serious occupational illness, such as cancer, asthma, asbestosis, silicosis, an illness related to chemical and fume inhalation, or permanent hearing loss	The program provides support for workers and their survivors in dealing with both the legal/policy and medical aspects of the occupational disease. Where a worker dies from the occupational disease, the program provides survivor benefits to qualifying survivors, including income replacement benefits, funeral costs, bereavement counselling, and assistance in joining the workforce.
Benefits for Seriously Injured Workers	Worker	Workers who have a permanent impairment and who receive NEL benefits assessed at 60% or more; who have a permanent impairment and receive permanent disability benefits equivalent to 100% of the maximum benefit level payable under the leglislation (the maximum benefit is 85% of either the worker's net annual earnings (NAE or "take-home pay") or 85% of the maximum compensable NAE (based on a gross earnings limit set annually, which is currently $73,300); or who have a "likely permanent" impairment that would warrant benefits in the ranges mentioned above, in the opinion of a WSIB health care professional.	Workers enrolled in this program are assigned a support team that includes a claims adjudicator, an occupational therapist, an advanced practice nurse case manager, and a nurse case manager. These professionals assist the worker with the adjustment to his or her disability and support the worker in living independently.

Most commonly, workers who attempt this kind of fraud do so without involving the employer in the wrongdoing. In these cases, the worker typically conceals the true details of his or her claim from both the employer and the WSIB.

Workers who are receiving WSIB benefits are required to report any "material change in circumstances" to the WSIB. The WSIB considers failure to report a material change in circumstances to be fraud, and such failure can have consequences for the worker (for example, a reduction in loss of earnings benefits to the 50 percent level, or even discontinuation of benefits). The WSIB provides the following examples of material change in circumstances:

- Changes in medical status or treatment—the worker becomes capable of doing modified work;

- Changes in earnings/income—the worker's wages from employment have increased or decreased or he begins to receive Canada Pension Plan disability benefits;

- Changes in work status—the worker has returned to work, has found a new job or a temporary part-time job, has been fired, or has retired;

- Changes in employment conditions—hours of work, wages, duties, and so on; and

- Situations that affect a worker's recovery or ability to work—for example, moving out of the province, going to prison, an unrelated health condition, or the expiry of a work visa.

Where the WSIB suspects that a worker has not reported a material change of circumstances, it may begin an investigation; as part of that investigation, it may contact the employer.

As the employer, it is important to be honest and straightforward in handling this kind of inquiry. However, you must also remember that you have obligations when it comes to protecting your employees' privacy, and that the information you recall about the case may not be up to date. In general, unless you are absolutely certain that your answers do not have privacy implications, you should respond to a WSIB request for information by making a note of the information requested and promising to respond as soon as you have checked your information.

Before responding, check the personnel file and speak with the employee's direct supervisors to determine the up-to-date status of the claim. Determine whether any of the information requested by the WSIB is personal information that you would need permission to disclose (check with your lawyer, if appropriate). Determine whether you have written consents on file, whether they are up to date, and whether they cover the request that was made; obtain new consents if necessary. Once you have made these inquiries, you will be in a better position to provide accurate and appropriate answers for the investigators.

Occasionally, management may be involved in a worker's fraud, usually by providing or allowing the worker to provide false information to the WSIB without attempting to correct this information. If you suspect that this might be the case at your business, contact a lawyer for assistance in responding to the investigation.

Employer Fraud and Non-Compliance

The most common kind of WSIA fraud committed by employers involves attempting to avoid paying all or part of a compulsory premium. Businesses may attempt to do this by, among other things:

- misrepresenting the nature of the business in an attempt to be classified in a less expensive rate group;

- not disclosing a particular business activity at all;

- underreporting payroll costs for the whole business;

- where a business falls into multiple rate groups, underreporting payroll for the higher rate group(s) and overreporting payroll in the lower rate group(s); and

- underreporting injuries and illnesses in an attempt to receive a discount or avoid a surcharge under an incentive program.

Employers are also in non-compliance with the WSIA when they fail to report incidents as required by the legislation. In a nutshell, employers must report within three days any time that, due to an incident at work:

- a worker loses work time;
- a worker earns less than a regular day's pay; or
- a worker requires medical treatment other than immediate first aid.

If a worker is hurt and needs only first aid, or is able to do modified work at regular pay, the employer has seven days in which to report the injury.

FACTS AND FIGURES

Employer Offences Under the WSIA

On its website, the WSIB gives the following as examples of activities that it considers "fraud or non-compliance":

- Failing to register within ten days of becoming an employer (WSIA section 151(1))
- Deducting workers' compensation expenses from workers' wages (WSIA section 155(2))
- Failing to notify WSIB within 3 days of learning of a worker's accident that requires health care or causes the worker to earn less than full wages (WSIA section 152(3))
- Failing to allow examination of payroll records by authorized persons attempting to confirm correct reporting of wages (WSIA section 153(1))
- Making a false statement or representation to the WSIB relating to benefit entitlement (WSIA section 149(1))
- Failing to inform the Board of a material change in circumstances within ten days (WSIA section 149(3))
- Failing to keep accurate records of all wages paid to employer's workers (WSIA section 152(1))
- Failing to produce records of all wages paid to employer workers (WSIA section 152(1.1))

If the WSIB suspects that a business is not in compliance with its obligations under the WSIA, it will typically refer the matter to its Regulatory Services Division.

In order to deter non-compliance, the WSIB administers an audit program. Businesses can be audited at random, or because there appears to be an irregularity in a business's reporting. Audits usually explore the current year plus two prior years; however, in some cases, an audit can go back five years. The WSIB provides general information on its website about how to prepare for an audit. Specific instructions are provided to individual audit subjects in advance.

Failure to comply with the WSIA can attract significant penalties. Individuals convicted of offences under the WSIA may be subject to fines of up to $25,000, six months' imprisonment, or both; corporations may be subject to fines of up to $100,000.

KEY TERMS

Workplace Safety and Insurance Act (WSIA)

premium

hearing

fraud

audit

REVIEW QUESTIONS

1. Where does the money held by the WSIA insurance fund come from?
2. Which employers are required to register with the WSIB, and how long do they have to register?
3. Identify at least one Schedule 1 class each of the following businesses would likely fall into:
    - a privately owned delivery company that employs a fleet of bicycle couriers
    - a furniture store
    - a hair and aesthetics salon
    - a greenhouse that sells plants directly to consumers
    - a business that distributes (but does not quarry) sand, gravel, patio stones, and other quarried materials on a wholesale basis to landscaping companies
4. Must a person meet the traditional employment law definition of permanent employee to be a "worker" for the purpose of the WSIA?
5. By providing no-fault benefits to injured workers, the WSIA system replaces the alternative—a system under which workers would have to take employers to court to sue for benefits under the law of tort. WSIA benefits are payable where an injury or illness is caused by an "accident." List at least three circumstances that would fall within the WSIA definition of "accident."
6. When a worker is diagnosed with an occupational illness, what is the "accident" date for the purpose of calculating benefits?
7. What is the relationship between CUs and rate groups under the WSIB's scheme of work classification?
8. What should an employer do if it believes that it has been classified incorrectly?
9. How (in a general sense) do the WSIB's NEER and CAD-7 incentive programs work?

10. The WSIB investigates cases of employer fraud and non-compliance every year. What are the maximum penalties chargeable for offences under the WSIA?

AT ISSUE: SHOULD THE WSIB PROVIDE COMPENSATION FOR CHRONIC PAIN?

A health problem is "chronic" if it persists over a long term and is resistant to treatment. Chronic pain afflicts some but not all workers who suffer compensable injuries under the WSIA. Chronic pain, including the reasons why some people heal better and feel less pain that others, is not well understood by modern medicine; nevertheless, it is well accepted, in the medical community, that many injuries lead to long-term, untreatable pain.

Because pain is hard to document and quantify, workers' compensation schemes have tended in the past to limit compensation on the basis of chronic pain. The Ontario WSIA once contained such a provision (section 14), but it has since been repealed (removed from the legislation). The repeal was at least partly in response to a Supreme Court of Canada decision that considered a similar provision in the Nova Scotia workers' compensation legislation. In that decision, the court said:

> By entirely excluding chronic pain from the application of the general compensation provisions of the [Nova Scotia] Act and limiting the applicable benefits to a four-week Functional Restoration Program for workers … the [Nova Scotia] Act and the FRP Regulations clearly impose differential treatment upon injured workers suffering from chronic pain on the basis of the nature of their physical disability, an enumerated ground under s. 15(1) of the *Charter*. In the context of the [Nova Scotia] Act, and given the nature of chronic pain, this differential treatment is discriminatory. It is discriminatory because it does not correspond to the actual needs and circumstances of injured workers suffering from chronic pain, who are deprived of any individual assessment of their needs and circumstances. Such workers are, instead, subject to uniform, limited benefits based on their presumed characteristics as a group. The scheme also ignores the needs of those workers who, despite treatment, remain permanently disabled by chronic pain … the denial of the reality of the pain suffered by the affected workers reinforces widespread negative assumptions held by employers, compensation officials and some members of the medical profession, and demeans the essential human dignity of chronic pain sufferers. The challenged provisions clearly violate s. 15(1) of the *Charter*.

Source: *Nova Scotia (Workers' Compensation Board) v. Martin; Nova Scotia (Workers' Compensation Board) v. Laseur* (2003).

Laws that are found by courts to violate a provision of the Charter can be stuck down and ruled to be of no force and effect. This is what happened to the Nova Scotia provision. While the Ontario provision provided only *limits* on compensation for chronic pain—instead of excluding it completely—the Ontario legislature chose to repeal the provision completely.

1. What do you think are some of the practical challenges associated with compensating workers for disabilities based on chronic pain?

2. Do you think that the potential for malingering or fraud on the part of workers is any greater in cases based on chronic pain than in other disability cases? Why or why not?

3. Do you agree with the Supreme Court's point that excluding claims based on chronic pain or applying a one-size-fits-all compensation formula to them "reinforces widespread negative assumptions" about injured workers' pain? Why or why not?

CHAPTER 6

Other Workplace Safety Laws

CHAPTER OBJECTIVES

After completing this chapter, you should be able to:

- List at least three subjects or issues regulated by the Industrial Establishments regulation.
- Understand when a Pre-Start Safety Review is needed at a workplace.
- Explain the difference between a legislated and a voluntary technical standard.
- List three sources of legislated technical standards and three sources of voluntary technical standards.
- Describe at least two human rights issues that have safety implications.
- Describe at least two kinds of workplace events that could trigger charges under the *Criminal Code*.

INTRODUCTION

As you learned in the introductory chapters of this book, worker health and safety has not always been considered a regulatory field of its own. Schemes designed to compensate workers who had been injured or sickened were the first to emerge. The comprehensive *preventive* approach prescribed by the OHSA, however, dates only to 1979.

Before 1979 the prevailing approach to worker safety involved including a provision or two on the subject in the general pieces of legislation created to regulate individual industries, processes, and classes of equipment. For example, before the OHSA was passed, the 1890 *Mining Operations Act* contained provisions related to safety issues, such as ventilation and lifting device standards.

While the OHSA now provides a general, multi-industry approach to safety, there are still safety-related legislative provisions in existence in a wide range of other statutes. In many cases, a statute does not have worker safety as its primary objective, but it addresses safety issues that are incidental to other objectives.

A good example of this kind of legislation are environmental laws. As you learned in chapter 2, the protection of the environment necessarily improves the safety of the people in that environment—including workers. For this reason, employers who are committed to safety must be careful to address not only the obligations imposed by the OHSA, but environmental compliance as well.

Two other general classes of legislation that have worker safety implications are human rights law and criminal law. Each of these is addressed in this chapter.

Finally, **technical standards**—standards that define appropriate design, engineering, performance, or other aspects of technologies and processes—also have an impact on worker safety. Technical standards fall into two broad classes: **legislated technical standards** (those that are contained in legislation and therefore have the force of law), and **voluntary technical standards** (those that are developed by the private sector). Both types of technical standards will be discussed here. First, however, we will look at the Industrial Establishments regulation, which for many workplaces is the most important regulation passed under the OHSA.

technical standard
an industry-recognized specification or set of specifications relating to a material, product, or service

legislated technical standard
a technical standard that is imposed and/or enforced by legislation or regulations

voluntary technical standard
a technical standard that is published or circulated within an industry as a suggested guidleline

THE INDUSTRIAL ESTABLISHMENTS REGULATION

Introduction

As you learned in chapter 3, there are many regulations "made under" the OHSA. While these are listed in appendix A, this text does not cover them in detail, because whether or not they apply depends a great deal on what kind of business an employer is running.

However, one regulation—the Industrial Establishments regulation—has wide enough application that it warrants mention here.

The Industrial Establishments regulation applies to "all industrial establishments," which are broadly defined, in the OHSA, to include: "an office building, factory, arena, shop or office" and any structures associated with one of these places. "Factory" also has a definition; it means a place where goods and articles are manufactured or made for sale. As you can imagine, a very large proportion of employers operate "industrial establishments" and will find themselves covered by this regulation. Certain operations, however, have separate schemes of regulation under the OHSA. Notable examples are mining, which is governed by the Mines and Mining Plants regulation, and health care, which is governed by the Health Care and Residential Facilities regulation. Employers who are uncertain about which OHSA regulations apply to them should contact an MOL inspector for guidance.

The Industrial Establishments regulation supplements the OHSA by providing detailed guidance on issues such as the following:

- minimum ages for logging, factory, and other workers;

- guidelines for the contents of the Notice of Accident reports prescribed by the OHSA under sections 51 and 52 (these were touched on in chapter 3 of this book and will be mentioned again in chapter 9);

- Pre-Start Health and Safety Reviews (discussed below); and

- specific safety guidelines for a wide range of workplace structural features, machinery, and processes.

FACTS AND FIGURES

Safety Specifications Under the Industrial Establishments Regulation

The Industrial Establishments regulation specifies minimum standards for a wide range of workplace structural features, machinery, equipment, and processes. These standards are designed to promote worker safety, and compliance with them is compulsory. Because these matters are covered in the regulation, employers can expect that they may be among the details checked during MOL inspections.

Voluntary standards for workplace equipment can also come from other sources. Some of these other sources will be discussed below. The standards included in the Industrial Establishments regulation are probably best viewed as a starting point for basic safety. Employers committed to excellence in workplace safety will want to investigate industry best practices to keep abreast of safety innovations as they emerge.

The Industrial Establishments regulation addresses the following:

- work floors, walkways, stairs, stiles, and access ladders;
- clearances (space) around machinery;
- guardrails;
- coverings for holes;
- labelling doors;
- warning signs and barriers to manage traffic;
- lighting;
- fire prevention and protection, including management of flammable liquids;
- machine guards, rests, locking mechanisms, and other features for preventing injuries;
- overhead protection (from falling objects);
- safety features and precautions for "explosive actuated fastening tools" (such as stud guns), nail guns, and chain saws;
- safety certification of electrical equipment, and general electrical safety;
- safe transportation and storage of materials to prevent tipping, falling, collapse burial hazards, and overhead hazards;
- safe use of cranes and hoists designed to lift workers, and the use of PPE in conjunction with this machinery;
- vehicle safety;
- labelling of pipelines and safety features of piping systems;
- ventilation and fume/dust management;
- safe management of "molten materials" and metal castings;
- safe handling of broken equipment and appropriate safeguards during repairs;
- guidelines for choosing and using PPE, such as safety glasses, helmets, shields, fall protection harnesses, flotation vests, and so on;
- training standards and certification for certain specific classes of workers, for example, cutter-skidder operators in logging;
- logging industry safety;
- guidelines for work done in confined spaces;
- industrial hygiene measures such as eye washing, deluge showers, and discharge of airborne contaminants;
- general environmental conditions management (working temperatures, access to drinking water, lunch rooms, and so on); and
- hearing protection and noise management.

Figure 6.1   Guidelines for Pre-Start Health and Safety Review, Section 7 Table

Item	Applicable Provisions of This Regulation	Circumstances
1	Sections 22(1), (2), and (4)	Flammable liquids are located or dispensed in a building, room, or area
2	Sections 24, 25, 26, 28, 31, and 32	Any of the following are used as protective elements in connection with an apparatus: • Safeguarding devices that signal the apparatus to stop, including but not limited to safety light curtains and screens, area scanning safeguarding systems, radio frequency systems and capacitance safeguarding systems, safety mat systems, two-hand control systems, two-hand tripping systems, and single or multiple beam systems • Barrier guards that use interlocking mechanical or electrical safeguarding devices
3	Section 45(b)	Materials, articles, or things are placed or stored on a structure that is a rack or stacking structure
4	Section 63	A process[a] involves a risk of ignition or explosion that creates a condition of imminent hazard to a person's health or safety.
5	Section 65	The use of a dust collector involves a risk of ignition or explosion that creates a condition of imminent hazard to a person's health or safety
6	Sections 87.3, 87.4, 87.5, 90(1), 90(2), 90(3), 91, 92, 94, 95, 96, 99, 101, and 102	A factory produces aluminum or steel or is a foundry that melts material or handles molten material
7	Sections 51 and 53	The construction, addition, installation, or modification relates to a lifting device, travelling crane, or automobile hoist
8	Sections 127 and 128	A process[a] uses or produces a substance that may result in the exposure of a worker in excess of any occupational exposure limit set out in Regulation 833, 835, 836, 837, 838, 839, 840, 841, 842, 843, 844, 845, or 846 of the Revised Regulations of Ontario, 1990

[a] For the purposes of this section, the term "process" refers only to those processes listed and identified in this table under items 4 and 8.

Source: Ontario Ministry of Labour, "Guidelines for Pre-Start Health and Safety Review: How to Apply Section 7 of the Regulation for Industrial Establishments," appendix I, http://www.labour.gov.on.ca/english/hs/guidelines/prestart/gl_psr_app1.html.

Pre-Start Health and Safety Reviews

A key compliance obligation created by the Industrial Establishments regulation is the (section 7) requirement that certain businesses or operations conduct a Pre-Start Health and Safety Review of their operations when certain changes are made. Determining precisely when a Pre-Start Health and Safety Review is needed is a fairly complicated exercise. The changes that trigger the need for a review are changes that affect the application of certain provisions of the regulation to the business. The relevant provisions are set out in figure 6.1, above. Changes that can prompt the need for a Pre-Start Health and Safety Review may be installations of, or modifications to apparatus, structures, protective elements, or processes.

A Pre-Start Health and Safety Review has two components:

1. **Review.** The employer must engage an expert (usually a professional engineer) to conduct a review of its business activities to assess the health and safety impact of any changes that trigger the obligation; to assess the safeguards and precautions that are in place, keeping in mind regulatory and industry standards; and to identify any needed modifications, improvements, or retrofit actions that would be necessary to achieve an adequate degree of safety for the workers.

2. **Report.** The expert must prepare a report detailing his or her findings. The expert must date and sign the report and the employer must store it, with any supporting documents, in a "readily accessible" location. Before the new apparatus or process is used, the report must be provided to the Health and Safety Representative or JHSC.

The MOL is currently developing a Frequently Asked Questions (FAQ) document covering the preparation of Pre-Start Health and Safety Reviews. In the interim, it has provided very general guidelines that can be accessed on its website at www.labour.gov.on.ca.

PROFILE

The Industrial Accident Prevention Association

The Industrial Accident Prevention Association (IAPA) is a not-for-profit member association, the largest of its kind in Canada.

On its website (at www.iapa.ca), the IAPA describes its mission as follows:

> To improve the quality of life in workplaces and communities we serve by being an internationally recognized leader in providing effective programs, products and services for the prevention of injury and illness.

The products and services provided by the IAPA include certification and training; safety consulting; conferences; and prose, audio, video, and other resources. The organization also sponsors charitable initiatives and awareness-raising campaigns, including the LifeQuilt memorial for young workers killed on the job and a young workers' safety-themed drama competition for students.

The IAPA has a membership of more than 50,000 businesses representing about 1.5 million workers from sectors including manufacturing and trades, high-tech, agri-business, government, and office services. Members enjoy basic consulting services, discounted pricing for IAPA products and services, and access to information and advice via forums such as the organization's Small Business Centre.

For employers, membership in the IAPA or in similar safety organizations is a way to stay on top of new developments in workplace health and safety and to benefit from peer best practices and knowledge. Membership and participation in training and certification programs also symbolizes an employer's commitment to being an industry safety leader.

A list of other important health and safety associations in addition to the IAPA appears in appendix B.

LEGISLATED TECHNICAL STANDARDS

Introduction

Technical standards for industrial equipment, machinery, processes, workspace layout, production engineering, and other workplace issues exist in legislation that does not have a specific safety focus. However, improved safety is often a by-product of adherence to technical industry standards, and so, compliance with all applicable legislation should be a priority for safety-conscious employers.

At the provincial level, examples of general technical standards laws with worker safety aspects and implications include:

- the *Technical Standards and Safety Act*;
- the *Building Code Act* and the *Building Code*;
- the *Fire Protection and Prevention Act, 1998* and the *Fire Code*;
- the *Electricity Act* and the *Electrical Safety Code*; and
- the *Dangerous Goods Transportation Act*.

Examples at the federal level include:

- the *Canada Oil and Gas Operations Act*;
- the *Explosives Act*;
- the *Hazardous Materials Information Review Act*;
- the *Hazardous Products Act*;
- the *Nuclear Safety and Control Act*;
- the *Canada Petroleum Resources Act*; and
- the *Transportation of Dangerous Goods Act, 1992*.

Employers should review these statutes to determine whether any apply to their business activities. While compliance with technical standards may not be a direct responsibility of managers charged with worker health and safety, it is important that these managers have a general understanding of the status of the organization's compliance with all relevant technical standards. A good compliance record will have important benefits from a safety perspective, and will make it easier to comply with safety-specific legislation such as the OHSA.

Overview of Legislated Technical Standards

THE TECHNICAL STANDARDS AND SAFETY ACT

The Ontario *Technical Standards and Safety Act* (TSSA) was proclaimed in 2001. It repealed a number of other pieces of subject-specific legislation (for example, the *Boiler and Pressure Vessels Act*) and "gathered" the standards that had been prescribed under those individual statutes into one coordinated legislative scheme.

The TSSA is administered by the Technical Standards and Safety Authority, a not-for-profit corporation (visit its website at www.tssa.org). The Authority performs training and certification functions for many businesses and professions, regulates areas of special concern (for example, amusement park rides, boilers/pressure vessels, and elevators), and provides useful resources to workers and managers in the regulated areas.

> ## Thinking Safe
> ### Lessons Learned—and Shared—by the Technical Standards and Safety Authority
>
> An example of the resources available from the Authority are the best practices protocols organized under the title "Lessons Learned." (See the TSSA website at www.tssa.org.) These protocols, or guidelines, describe a technical problem or equipment failure encountered by inspectors or business owners. They also describe concrete ways—equipment choices, safety pratice recommendations, and process changes—to prevent the problem from arising in the future.
>
> The following is an example of the content of one of these resources. The lesson entitled "Potential Failures of Carabiners on Amusement Device" explains that inspections of an amusement park ride known as the Euro Bungee have revealed that ball-lock-style carabiners (locking hooks) can sometimes fail in the open position. The best practices guideline goes on to suggest specifications for carabiners (for example, choose screw lock type, minimum breaking strength of 3,000 kilograms, and so on) that should prevent carabiner failure in the future.
>
> Making these kinds of recommendations public allows other businesses to benefit from the safety experiences of their peers. Where a guideline emerges in the wake of an accident, publication of a best practice ensures that some good will come out of the situation.

In addition to providing its own technical guidelines for certain aspects of the equipment and processes it regulates, the Authority also has the power to "adopt" (essentially, give the force of law to) non-legislated codes of standards. Many (though not all) of the codes adopted by the TSSA were developed by the Canadian Standards Association (CSA). Examples of codes adopted by the TSSA include:

- CSA W47.1-1992, Certification of Companies for Fusion Welding of Steel Structures
- ANSI A10.22-1990 (R1998), Safety Requirements for Rope-guided and Non-guided Workers Hoist
- CSA B149.1-00, Natural Gas and Propane Installation Code

HARD LESSONS
Fines Flow Under TSSA from Work on Elevator by Unregistered Contractors

One of the services provided by the Technical Standards and Safety Authority is the certification of tradespeople. In many cases, legislation or regulations require that certain work be performed only by individuals possessing certification that reflects certain standards of training in a particular task.

> For example, a regulation made under the TSSA requires that work on elevators be conducted only by, or under the direct supervision of, registered contractors who possess at least one of nine available certifications for work done on elevators. This requirement is designed to ensure the safety of the public by requiring that people who alter or repair elevators have appropriate expertise.
>
> A January 2007 media release reported that two Whitby, Ontario contractors had pled guilty to offences under the TSSA. The workers pled guilty to altering an elevator without the appropriate registration, and were also charged with failing to register a design submission for the elevator, failing to obtain an inspection of the elevator before allowing it to be operated, and operating an unsafe elevator. They were fined $3,000 each.
>
> The contractors were also assessed a 25 percent "victim surcharge." For more information about victim surcharges, see the Thinking Safe feature later in this chapter.
>
> In its media release about the case, the Technical Standards and Safety Authority expressed a willingness to fine individuals in appropriate cases, explaining:
>
> > Bringing charges against companies or individuals for performing unregistered work, using unapproved or unlicensed equipment, or for failing to follow proper procedures and regulations, is part of TSSA's safety mandate, and strongly reinforces prosecution objectives—to deter violators and increase public safety.

BUILDING CODES AND FIRE CODES

Building and fire codes exist in jurisdictions across Canada. These codes create standards that typically have the force of law (most building and fire codes are passed "under" authorizing legislation).

The main purpose of building and fire code standards is the protection of the public. The specific provisions of these codes are based on experience with structural and fire hazards, and design and safety features designed to minimize those hazards.

The Ontario *Building Code* is a regulation made under the *Building Code Act*. Based on the *Building Code* of Canada, it establishes standards for the construction of buildings. These standards are updated frequently to take into account new expertise and to harmonize compliance with other legislation. For example, 2003 updates to the *Building Code* ensured that it incorporated drinking water protection guidelines established under the *Safe Drinking Water Act, 2002*. The *Building Code* is an extremely long and complex regulation, and is merely introduced here. Full compliance will require a careful study of the Code in the context of the employer's specific situation in terms of physical facilities and business type.

When a building is constructed, it must, before being occupied, be inspected and found to comply with current building code standards. This means that older buildings are not necessarily in compliance with the current version of the Code (however, any renovations made to those buildings must comply with current standards).

The *Building Code* classifies buildings according to type of occupancy. The general classification scheme is as follows:

Group A: Assembly occupancies
Group B: Care or detention occupancies
Group C: Residential occupancies
Group D: Business and personal services occupancies

Group E: Mercantile occupancies
Group F: Industrial occupancies

Group F is subdivided into three classes: 1—high hazard, 2—medium hazard, and 3—low hazard occupancies.

The first few provisions of the *Building Code* provide a detailed explanation of which sections of the Code apply to occupancies within each group. Factors such as size and height of a building, in storeys, can dictate the application of certain provisions, as can specifics with respect to business activities.

The kinds of structural and design features that are regulated by building codes include:

- building materials;
- heating, plumbing, sewage, ventilation, and outdoor lighting systems;
- firewalls, fire resistance ratings, fire protection systems;
- the number of bathrooms and toilets; and
- the number and placement of exit doors, door access, and door features.

The above list is merely a sampling, and is not exhaustive.

Businesses moving into new buildings would be well advised to confirm that the appropriate building code inspections were performed at the time of the building's construction and at the time of any renovations. If the employer has plans to renovate or retrofit the building for its own purposes, compliance with the *Building Code* is essential, especially where there will be a change in occupancy type. Contraventions of the Ontario *Building Code Act* can receive fines of up to $100,000 for individuals, and $200,000 for corporations.

Fire codes, like building codes, are generally established as regulations under a parent statute; in Ontario, the *Fire Code* is a regulation made under the *Fire Protection and Prevention Act, 1997*. The *Fire Code* is administered by the Ontario Fire Marshal (www.ofm.gov.on.ca). The party with primary responsibility for compliance with the *Fire Code* is the building *owner*. This means that where an employer leases commercial or industrial facilities from a landlord, the employer must rely on the landlord to comply with the Code. Where an employer who does not own a building has concerns about fire safety, the employer should raise these concerns with the building owner first. If the owner seems unwilling to comply, the employer may then have to take further steps—for example, reporting the problem to the fire marshal—as a part of its duty to protect the safety of workers.

The *Fire Code* uses the same occupancy classifications as the Ontario *Building Code*. Like the *Building Code*, its primary objective is the promotion of public safety; however, there is a narrower focus on fire, smoke, and fire-related issues. Also like the *Building Code* and the TSSA, the *Fire Code* adopts a range of existing safety, design, and construction standards into its own regulatory scheme.

One of the requirements of the *Fire Code* that is probably most familiar to employers is mandatory testing of fire protection systems (for example, alarms) in certain occupancies. All building owners must have a fire safety plan for each building, and this plan should include:

- emergency procedures in case of fire (these will include an evacuation plan and procedures for controlling and/or extinguishing the fire);

- designation of staff responsible for supervising emergency procedures;
- instruction in fire safety procedures, especially for staff designated as supervisors for these procedures; and
- a regular program of fire drills.

Part 4 of the Code includes specific guidelines for the handling of flammable and combustible liquids, while part 5 includes guidelines for flammable, combustible, or explosive non-liquids (for example, demolition explosives, fireworks, compressed gas cylinders, and flammable plastics like those made of cellulose nitrate). Employers who use, manufacture, and/or store these materials should study the *Fire Code* carefully to ensure compliance. Guidelines for managing these materials exist under other legislation as well, for example, in the Industrial Establishments regulation made under the Ontario OHSA; under legislation dealing with the transportation of dangerous goods; and under the WHMIS.

SAFETY IN THE USE OF RESOURCES—WATER, GAS, AND ELECTRICITY

Another group of statutes deals with the safe production and management of resources such as electricity, petroleum products, water, and nuclear energy. These resources are subject to government regulation to ensure fairness and safety. Businesses that produce, refine, supply, or otherwise handle these resources will certainly need to be aware of their obligations under these statutes; however, managers with responsibility for technical compliance may benefit from being encouraged to strive for the highest standards of performance, in their areas of responsibility, in the name of safety.

Compliance with the regulation of resources is typically overseen by governmental and quasi-governmental agencies, for example:

- The (Ontario) Electrical Safety Authority (www.esa-safe.com);
- The Canadian Nuclear Safety Commission (www.nuclearsafety.gc.ca);
- Emergency Management Ontario;
- The Nuclear Waste Management Organization (www.nwmo.ca); and
- The National Energy Board (www.neb.gc.ca).

These agencies provide not only regulatory, but also support services to businesses involved in the management and supply of resources such as electricity; nuclear power; and oil, gas, and petroleum products.

DANGEROUS GOODS AND MATERIALS

Several statutes, at both the provincial and federal levels, deal with the management of hazardous products and materials. These statutes overlap, to some extent, with certain provisions of the OHSA and the regulations made under that Act, many of which regulate specific hazards (Regulation 842, for example, regulates isocyanates in the workplace). Management of hazardous products within the workplace was discussed in detail in chapter 4 of this book.

Where hazardous products are transported from one location to another, separate regulatory schemes apply. The *Transportation of Dangerous Goods Act, 1992* (TDG), a federal statute, creates a comprehensive scheme for the regulation of dangerous goods in transit. The regulations made under the TDG have been "adopted" by each province (in Ontario, via the *Dangerous Goods Transportation Act*). These adoptions were accomplished to create a harmonized scheme (similar to the WHMIS) and to eliminate jurisdictional complications related to the constitutional status of the subject matter (as was discussed in chapter 2, the regulation of trade and commerce within a province is an area of provincial jurisdiction).

The scheme for classifying hazardous goods is based on international standards—a choice designed to simplify the management of imports of goods into Canada. There are nine general classes of dangerous goods:

Class 1: Explosives
Class 2: Gases (compressed, deeply refrigerated, liquefied, or dissolved under pressure)
Class 3: Flammable and combustible liquids
Class 4: Flammable solids (substances liable to spontaneous combustion or that emit flammable gases on contact with water)
Class 5: Oxidizing substances; organic peroxides
Class 6: Poisonous and infective substances
Class 7: Radioactive materials
Class 8: Corrosive agents
Class 9: Miscellaneous products and substances that fall into none of the other eight categories but that are nonetheless dangerous to health or life when transported.

The TDG, its regulations, and the related provincial legislation and regulations govern every stage of transportation, including packing, unpacking, and storage in transit. In many cases, this legislation prescribes specific packaging for dangerous goods. It also prescribes standards for labels and placards to warn of the dangers of what is being transported, and requires that the transporter carry and have available prescribed information sheets describing the contents and explaining the properties of the cargo. While goods are marked in conformance with the TDG, the requirements of the WHMIS (WHMIS packaging and information requirements are somewhat different) are waived; however, once goods are unloaded at the workplace, the employer must immediately take steps to bring all packaging, labelling, and information resources into conformity with the WHMIS.

Finally, personnel who transport dangerous goods must meet prescribed standards of training for this task. Training standards under the TDG are, again, slightly different from WHMIS materials handling standards.

TRAINING AND CERTIFICATION STANDARDS

Ontario's *Apprenticeship and Certification Act, 1998* was created with the following objectives:

- to support and regulate the acquisition of skills for trades and other occupations through workplace-based apprenticeship programs that lead to formal certification;

- to promote quality training for trades and other occupations; and

- by the means set out in clauses (a) and (b), to expand opportunities for Ontario workers, increase the competitiveness of Ontario businesses, and ensure public and worker protection.

As you can see, safety—in the form of "public and worker protection"—is a central objective of training standards for tradespeople and workers serving in "other occupations and skill sets." The *Apprenticeship and Certification Act* has an older, sister statute: the *Trades Qualification and Apprenticeship Act*. Both pieces of legislation are similar in that they create a framework for the supervision of apprenticeships and programs of testing and certification. Where they differ is in the trades and occupations they govern.

The *Apprenticeship and Certification Act* governs a wide range of general occupations, as well as a list of "restricted skill sets," most of which relate to the repair of passenger vehicles. The *Trades Qualification and Apprenticeship Act* is designed to govern a range of trades, mostly in the construction industry, which are *not* covered by the *Apprenticeship Certification Act*. The specifics of which trades are covered by which statute can be determined by reference to the regulations made under each.

Both statutes are administered by Ontario Skills Development, a governmental agency within the Ministry of Training, Colleges and Universities. The *Apprenticeship and Certification Act* creates the office of Director of Apprenticeship. The Director oversees the creation of standards-setting committees and awards certifications under both statutes.

From the employer's perspective, a commitment to worker safety requires that any work done by the business that falls within a regulated trade must be done by an appropriately certified person. Section 13 of the *Apprenticeship and Certification Act* provides:

> A person shall not employ or otherwise engage an individual to perform a skill that is part of a restricted skill set unless the individual is authorized to perform that skill.

Contravening this section can, upon conviction, form the basis of a fine of up to $25,000. Subsection 10(3) of the *Trades Qualification and Apprenticeship Act* provides:

> No person shall employ any person, other than an apprentice or a person of a class that is exempt from this section or a person referred to in subsection (4), in a certified trade unless the person employed holds a subsisting certificate of qualification in the certified trade.

(Workers "referred to in subsection (4)" are workers who were already working in the trade before it became a certified trade; these workers can continue working while they apply for the new certification based on their existing qualifications.) Employers who contravene this section can face fines of up to $2,000.

VOLUNTARY STANDARDS

Introduction

Statutes are drafted by politicians and lawyers, not by tradespeople and engineers. While some technical standards are given the force of law by legislatures, most standards do not originate in the offices and meeting rooms of parliamentarians. Instead, they are developed on the job, fine-tuned by experts in complex areas of speciality, and disseminated to other businesses by industry leaders.

Sometimes, this is an informal process: an industry leader discovers a way of doing something that promotes safety, enhances productivity, or both, and the innovation is soon copied by competitors.

In other cases, standards are developed and disseminated through a more formal process that includes the sanction of a recognized standard-setting organization. Standards developed in this way lack the force of law (at first, anyway—see the next paragraph), but they are often readily accepted and followed by employers eager to improve their performance and to be recognized as leaders in the industry.

In recognition of the safety expertise of industry leaders, it has become common practice in Canada for parliaments at both the federal and provincial levels to "adopt" certain standards developed by the private sector. This adoption, which comes typically via reference in a statute or regulation, effectively gives once-voluntary industry standards the force of law.

Examples of this process have already been described in this text. In chapter 3, we mentioned that section 40 of the Industrial Establishments regulation made under the OHSA requires that electrical equipment in use in an establishment governed by the regulation be certified by either the CSA or the Electrical Safety Authority; in this chapter, above, we noted that the *Technical Standards and Safety Act* adopts a wide range of technical standards that originated in the private sector.

Sources of Industrial Safety Standards

A number of organizations set voluntary safety-related standards. These organizations exist at the international, national, and provincial levels, and can be either general or trade-specific in application. Employers going into business should familiarize themselves with the standards-setting organizations available for membership, and should consider joining. Membership in these associations can be an excellent source of support to an employer's safety initiatives and IRS. A good place to start learning about standard-setting organizations is the Standards Council of Canada (www.scc.ca).

Below are some examples of standard-setting organizations.

At the international level:

- the International Standards Organization (ISO)
- the International Electrotechnical Commission (IEC)
- the International Atomic Energy Agency (IAEA)
- the International Association for Food Protection (IAMFES)

At the Canadian federal level:

- the Canadian Standards Association (CSA)
- the Canadian General Standard Board (CGSB)
- the Canadian Centre for Occupational Health and Safety (CCOHS)
- the Underwriters' Laboratories of Canada (ULC)

At the US federal level:

- the National Institute for Occupational Safety and Health (NIOSH)
- the American National Standards Institute (ANSI)

At the Canadian provincial level:

- the Ontario Petroleum Contractors Association
- the Nova Scotia Dental Association
- the Association of Manitoba Land Surveyors
- the Association of Professional Engineers and Geoscientists of British Columbia

An excellent example of an important "voluntary" safety standard is the Canadian Standards Association's standard number CSA-Z1000, Occupational Health and Safety Management. This standard will be discussed in detail in chapter 8.

FACTS AND FIGURES
Other Useful CSA Standards

Besides CSA-Z1000-06, Occupational Health and Safety Management, the CSA identifies the following standards as also relevant to occupational health and safety. You may wish to obtain copies of the relevant ones for your occupational health and safety resource collection:

- Z432-04, Safeguarding of Machinery
- Z460-05, Control of Hazardous Energy—Lockout and Other Methods
- CAN/CSA-Z731-03, Emergency Preparedness and Response
- CAN/CSA-Z796-98 (R2003), Accident Information
- CAN/CSA-ISO 19011:03, Guidelines for Quality and/or Environmental Management Systems Auditing

SUBJECT-SPECIFIC LAWS WITH SAFETY ASPECTS

Employers need to be aware that even if the main focus of a statute is not worker safety, there are provisions related to safety in a wide range of subject-specific statutes; that is, statutes that regulate a specific activity, substance, class of equipment, profession, and so on. Examples of subject-specific statutes that contain at least one safety-related provision are provided below.

At the Ontario provincial level:

- the *Day Nurseries Act*
- the *Farm Implements Act*
- the *Healing Arts Radiation Protection Act*
- the *Health Protection and Promotion Act*
- the *Mandatory Blood Testing Act*
- the *Mining Act*
- the *Pesticides Act*

At the federal level:

- the *Aeronautics Act*
- the *Canada Shipping Act, 2001*
- the *Canada Labour Code*
- the *Canadian Transportation Accident Investigation and Safety Board Act*
- the *Corrections and Conditional Release Act*
- the *Department of Health Act*
- the *Non-Smokers' Health Act*
- the *Pest Control Products Act*
- the *Railway Safety Act*

The *Healing Arts Radiation Protection Act*, as an example, is designed to protect everybody—not just workers, but also the public—from hazards associated with exposure to X-ray technology. The legislation and the *X-Ray Safety Code* regulation made under it accomplish this by describing standards for X-ray equipment; its installation, licensing, and registration; and the necessary qualifications for operators. The legislation also makes it compulsory to comply with structural shielding provisions and the wearing of personal protective equipment (PPE). These provisions in particular are aimed at ensuring worker safety, and if X-ray equipment is used in a business's operations, compliance with them is an essential element of overall occupational health and safety (OHS) compliance.

Non-compliance with a provision of a statute can often form the basis for the charging of an offence. Sometimes, the penalties chargeable and the procedure for prosecuting offences are set out in the offence-creating statute itself. In other cases (for example, where a statute creates offences but does not describe penalties or procedures), these are governed by the provisions of the *Provincial Offences Act* (POA), an Ontario statute of general application that guides the prosecution of provincial offences (that is, offences created by statutes passed by the Ontario legislature).

Thinking Safe

Victim Fine Surcharges

Non-compliance with a provision in a provincial statute can sometimes form the basis of an offence. Many provincial offences are governed by the *Provincial Offences Act* (POA).

Where an incident of non-compliance results in injury or other harm to a victim (in contrast to, for example, cases in which non-compliance is revealed in the course of a routine inspection), section 60.1 of the POA allows the court to impose a "victim fine surcharge" in addition to the fine assessed. This surcharge is designed to reflect the increased gravity of a violation that results in an actual safety breach and harm.

Victim fine surcharges as a flat fee on fines of up to $1,000 (the surcharge on a fine of $1,000 is currently $125). For fines over $1,000, the surcharge is 25 percent of the total fine.

HUMAN RIGHTS LAW

Introduction

While human rights legislation might not immediately come to mind when thinking about worker safety, respecting human rights is a key component of health and safety compliance. Canada has a diverse population, and employers are not permitted to discriminate—in their hiring and other employment practices—against individuals on the basis of their individual characteristics. Employers are also prohibited from tolerating discrimination and harassment among co-workers. But how does this relate to occupational health and safety? Consider the following scenarios:

- A candidate for a job is asked, during her interview, whether she has ever made a claim under the WSIA. She answers yes, and doesn't get the job.
- A rumour begins circulating among the staff of a security company that a particular employee is gay. Even though security guards are expected to do their patrol rounds in pairs (for safety's sake), this employee's co-workers refuse to accompany him on patrol. While doing rounds alone, he surprises an intruder and is injured in a struggle.
- A manufacturing company with a mostly male workforce hires a female employee to work on the factory floor. Her co-workers resent her presence and she faces an escalating campaign of sexual harassment that provokes a serious depression and a mental-health-based claim under the WSIA.
- A landscaping company employee suffers from significant obesity related to the side-effects of a prescription drug. One day when the company is doing work outside in extremely hot weather, this employee exercises his right to refuse unsafe work, alleging that his obesity is a disability that makes him hypersensitive to extreme temperatures.

As you can see, human rights and safety are indeed related issues in the workplace.

Human rights legislation exists at both the provincial and the federal levels. While the protection of human rights is identified by the Charter as a sphere of *provincial* responsibility, the federal government regulates the guarantee of human rights "within the purview of matters coming within the legislative authority of [the federal] Parliament." In practice, however, because most of the contexts (employment, commerce, access to housing, access to government services) in which human rights conflicts arise fall within the provincial sphere of authority, the provincial human rights codes are much more frequently engaged (in Ontario, the *Human Rights Code*).

The most important function of human rights legislation is to provide recourse for victims of discrimination. The *Canadian Human Rights Act* expresses the following as prohibited grounds of discrimination:

> race, national or ethnic origin, colour, religion, age, sex, sexual orientation, marital status, family status, disability or conviction for an offence for which a pardon has been granted.

The Ontario *Human Rights Code*, in turn, lists the following prohibited grounds:

> race, ancestry, place of origin, colour, ethnic origin, citizenship, creed, sex, sexual orientation, age, marital status, family status or disability.

Safety-Related Human Rights Issues

According to the Ontario Human Rights Commission (OHRC), approximately three-quarters of all human rights complaints come from workplaces. As the examples provided above illustrate, many human rights issues have a safety component. The following is a brief introduction to some of the issues employers should consider when thinking about human rights and safety.

DISCRIMINATION IN HIRING

Employers are not allowed to discriminate, in their hiring, on the basis of the grounds described in human rights statutes. Where an employee has a disability or handicap, the employer can neither refuse to hire nor fire that person on the basis of the disability unless the ability impaired by the handicap is essential to the work. For example, in certain rare cases (for example, aviation), excellent eyesight may be a legitimate work requirement. In the vast majority of cases, however, vision that is less than 20/20 should not preclude a candidate from getting a job.

A more subtle way in which some interviewers may screen out "disabled" candidates is by asking questions about whether candidates have ever made a claim under the WSIA. This kind of questioning constitutes discrimination, and could form the basis of a human rights complaint.

If an employee becomes disabled while on the job, the employer has a duty to accommodate him or her, for example, by changing the way the work is done so that the employee is able to do it. A simple example of accommodation might be installing a flashing light to supplement the audible ringer on an office phone so that hearing-impaired employees will know when the phone is ringing.

SAFETY RISKS HEIGHTENED BY, OR RELATED TO, DISABILITY OR DIFFERENCE

As noted in two of the examples above (those of the gay security guard and the obese landscaper), individual characteristics can influence a worker's risk of injury or illness at work.

It is easy to understand how a person with a disability might have more difficulty than average while navigating the workplace landscape. As you will see below, in the section that discusses the *Ontarians with Disabilities Act*, some employers have a legal obligation to make their workplaces accessible to adults with disabilities. However, even employers who do not have obligations under that legislation have a general duty to protect the safety of their employees. These employers, though they may not be required to build ramps or install grab bars, will still need to pay close attention to the risks faced by workers in the workplace. In the obese landscaper scenario, for example, a prudent employer should not challenge the employee's work refusal on the grounds that he is the only employee refusing work. Workers are entitled to refuse work that threatens their own *individual* safety—with all disabilities and individual characteristics considered—and not just work that seems unsafe from a general or majority perspective.

Similarly, as highlighted in the gay security guard scenario above, employers have a duty to pay attention to the interpersonal dynamics in the workplace so that they can intervene to prevent harassment or mistreatment on the basis of a characteristic that is recognized as grounds for discrimination.

Failure on the part of an employer to accommodate disabilities or to enforce respect for human rights can form the basis of a human rights complaint.

HEALTH SCREENING, SUBSTANCE-USE TESTING, AND COMPULSORY IMMUNIZATION

Another issue with both human rights and safety aspects is the employer's "intrusion" into the health or health habits of employees. Health screening, substance-use testing, and compulsory immunization are all issues that fall into this category.

In a few professions, it is considered to be prudent, from a safety point of view, to screen existing and potential employees for particular health problems. For example, astronauts in training are typically subjected to tests designed to measure their physical tolerance of the unusual forces (for example, G-forces) to which they might be subject in space. The point of this testing is to identify individuals who might, once blasted into space, find themselves too incapacitated by symptoms to discharge their job duties. A more down-to-earth example is the testing of educators for immunity to tuberculosis, a contagious disease that may be encountered or spread on the job.

However, it is generally considered to be an invasion of personal privacy—and therefore inappropriate—to subject employees or potential employees to medical screening. Perhaps more importantly, it is a practice that opens the door to potential discrimination on the basis of "disability" broadly defined. Where the need to screen based on health is essential to the job, the kind of screening permitted is described in statutes. An example is section 35 of the *Railway Safety Act*, which provides for regular medical examinations of individuals in specially designated positions considered to be "critical to safe railway operations." Employers whose work is not governed by these special provisions will not generally be permitted to require that candidates submit to medical examinations.

The OHRC's general policy on requests for medical information in the hiring process is available from the Commission, or on its website at www.ohrc.on.ca. Highlights of the policy include the following:

- It is appropriate to state *essential* health-related requirements in job postings (this is done for the benefit of applicants, so that they do not waste time applying if they do not meet the requirements), but it is not appropriate to ask questions about health status on an application form.

- If a medical examination is warranted as part of the application process, it should not be required until *after* the candidate has been made a conditional offer of employment.

- In a personal interview, it is acceptable for employers to ask candidates if they have any disabilities that might require accommodation on the job; other health inquiries are generally not appropriate.

A related issue is drug and alcohol-use screening. While impairment due to substance abuse can indeed increase the risk of an accident, *addiction* to alcohol or drugs is actually protected as a "disability" under the Ontario *Human Rights Code*. This means that it is not appropriate to fire, refuse to promote, or refuse to hire a person because he or she has a substance addiction.

Having a substance addiction and using or being impaired by a substance *on the job* are two separate issues, however. Employers have a duty to protect the safety of *all* employees, and so they are not usually required to tolerate on-the-job substance use or impairment. Detecting—and proving—impairment can be difficult without the use of testing, such as urine testing. Also, substance testing may reveal *past* substance use that is unrelated to on-the-job impairment (for example, a test may reveal traces of marijuana that relate to off-the-job use several weeks prior to the test). These kinds of results have the potential to open the door to discrimination even where the test subject has never been impaired on the job. The OHRC has a written policy governing drug and alcohol testing. The policy begins with the presumption that testing is a discriminatory practice, but that it is permissible on an exceptional basis once a certain threshold of criteria is met.

Highlights of the policy (which can be viewed on the OHRC's website) appear below.

1. Testing is only appropriate where the employer has met the following three-part test:

   - the employer has adopted the standard or test for a purpose that is rationally connected to the performance of the job;

   - the employer adopted the particular standard or test in an honest and good faith belief that it was necessary to the fulfillment of that legitimate work-related purpose; and

   - the standard or test is reasonably necessary to the accomplishment of that legitimate work-related purpose. To show that the standard is reasonably necessary, it must be demonstrated that it is impossible to accommodate individual employees sharing the characteristics of the claimant without imposing undue hardship upon the employer.

2. An employer's testing policy should be applied in a way that guarantees equality. Employers should not "single out" some employees and not others, for example, if an employer responds to an incident of suspected impairment by testing in one case, it should test in every case.

3. Testing should focus on measuring impairment, not past use, and even then is only appropriate where impairment is related to performance (for example, the individual being tested is in a safety-sensitive role, such as a school bus driver).

4. Because pre-employment drug and alcohol testing necessarily measures only substance use outside the workplace (because the person has not been hired yet), it is not likely jusitifiable as part of any pre-employment medical testing.

The final issue in this cluster is compulsory immunization. While immunization can reduce the risk of contracting (and spreading) certain diseases, immunization carries risks of its own. An employer's requirement that an employee be immunized against something, therefore, is problematic from a legal and a human rights point of view.

In Canada, laws exist that require schoolchildren to be immunized against certain diseases (for example, polio, pertussis, diptheria, tetanus, measles, mumps, and rubella) or to have a valid excuse for not being vaccinated (for example, compromised immune status). These laws have been around a long time, and the practical result is that most Canadian adult employees are vaccinated against these diseases.

Other vaccines, however, have emerged more recently. Not all adult Canadians, for instance, have been vaccinated against hepatitis B, a virus that is fairly easy to catch, especially when working in a health care setting. The hepatitis B vaccine has been shown to be quite effective, and so many health care workers receive it voluntarily, without being required to do so. Requiring a worker (for example, a bookkeeper) who has a *low* workplace risk of exposure to the virus, however, would be legally problematic.

One vaccine that employers have begun to focus on is the influenza (flu) vaccine. Unlike other disease-specific vaccines (such as the chicken pox vaccine), the flu vaccine does not provide the person who receives it with a particularly high likelihood of escaping the flu. This is because there are many different strains of flu, and it is possible to have immunity against one and not others. Every year, the makers of the flu vaccine attempt to predict which strains of the flu will be most prevalent in the coming season and prepare a vaccine to match. The predictions are not always accurate, however, and the duration of immunity tends to be brief—several weeks or a few months.

Nevertheless, since employers suffer from the loss of employees every year from flu, it can be tempting to attempt to require immunization, especially in the context of widespread talk of particularly dangerous strains (for example, "avian flu") or the likelihood that Canada is "due" for a flu pandemic.

At the moment, there is no legal general basis for requiring employees to be immunized against diseases. Individual employers of employees at special risk of things like hepatitis or flu sometimes develop policies encouraging employees to be immunized, for example, at on-the-job clinics. However, it is very likely that refusal to hire a person, firing a person, or restricting the job duties of a person who opts *not* to be immunized would amount to discrimination under human rights legislation.

CO-WORKER HARASSMENT

Harassment can affect health and safety in different ways. If the harassment escalates to the point of physical violence, sabotaging machinery with the intent to cause injury, or sexual assault, the risks are obvious. However, even verbal harassment can contribute to the development of mental health problems and psychophysiologic illness (illness that results from the processes of the mind acting on the body).

Employers must be alert to the potential for employees to harass their co-workers, and knowingly permitting such harassment often amounts to discrimination under human rights statutes. Employers should work to foster an environment in which discrimination and harassment are never tolerated, and in which employees are made to feel as comfortable as possible about speaking with supervisors about their concerns. If harassment does arise, the employer should move swiftly to denounce it, which often means disciplining the harasser. Persistent harassers who do not respond to discipline must be fired. In his book *Work Rage: Identify the Problems, Implement the Solutions*, Gerry Smith notes:

It could be argued that a racially motivated act of violence—say, a physical assault—as well as being criminally wrong, is an act of harassment. An employer would have to be careful to ensure that protective policies and procedures are in place if she or he wanted to avoid the costs associated with successful complaints under human rights legislation. (Smith, 2000, p. 133)

As the examples at the beginning of this section illustrate, discrimination by employees need not escalate to the point of harassment to constitute harm. Passive discrimination—for example, refusing to cooperate with a co-worker on the basis of a recognized ground of discrimination—can be dangerous when cooperation is essential to safety.

There are many useful resources available that can help an employer prevent harassment on the job. One such resource is the Canadian Human Rights Commission's *Anti-Harassment Policies for the Workplace: An Employer's Guide*. This document, as well as anti-harassment policy checklists for small and medium-to-large organizations, is available on the Commission's website at www.chrc-ccdp.ca.

Ontarians with Disabilities Act

Another key piece of legislation in the area of human rights is the *Ontarians with Disabilities Act* (OWDA). This statute imposes obligations on certain employers to take steps to make their workplaces accessible to people with disabilities.

The employers covered under the OWDA include the government of Ontario, municipal government employers, public transportation organizations, "educational institutions," public hospitals, and certain other agencies that receive capital funding from the government. The legislation also includes a definition of "agency" that would permit cabinet to list other agencies in regulations made under the Act (there are, as of yet, no such regulations).

The OWDA requires employers to take steps to remove not only physical barriers to accessibility, but other kinds of barriers as well. A "barrier" is broadly defined in the legislation as:

> anything that prevents a person with a disability from fully participating in all aspects of society because of his or her disability, including a physical barrier, an architectural barrier, an information or communications barrier, an attitudinal barrier, a technological barrier, a policy or a practice.

While the mandatory application of the OWDA is restricted to the public and quasi-public sectors, the legislation encourages the promotion of accessibility in the private sector as well. The preamble (introduction) to the OWDA points out that two statutes—the *Corporations Tax Act* and the *Income Tax Act*—create credits for employers to help offset the costs of renovations and modifications designed to improve accessibility.

THE CRIMINAL LAW

Introduction

Last but not least on the list of statutes that play a role in regulating worker safety is the *Criminal Code* of Canada. The *Criminal Code* applies to workplace safety in at least two key ways: as a tool for responding to workplace violence, and as a source of law that regulates acts of negligence so serious that they are deemed to be criminal in character.

Along with the *Criminal Code*, there are other statutes that make up the general criminal law; these include the *Victims' Bill of Rights* and the *Criminal Injuries Compensation Act*. While a discussion of these is beyond the scope of this text, it is useful to know that they address victims of crimes, their right to participate in the criminal justice system, and their access (in some cases) to compensation for criminal injuries.

Workplace Violence

Violence has always been an unfortunate feature of many workplaces. Acts of violence or sexual assault can occur between employees and their supervisors, or between employees and their co-workers, clients, family members, or other workplace outsiders.

While most people think of physical assaults when they think of "violence," the CCOHS provides a much broader definition of workplace violence (on its website) that includes the following:

- threatening behaviour—such as shaking fists, destroying property, or throwing objects

- verbal or written threats—any expression of an intent to inflict harm

- harassment—any behaviour that demeans, embarrasses, humiliates, annoys, alarms, or verbally abuses a person and that is known or would be expected to be unwelcome. This includes words, gestures, intimidation, bullying, or other inappropriate activities.

- verbal abuse—swearing, insults, or condescending language

- physical attacks—hitting, shoving, pushing, or kicking

When serious violence or sexual crime erupts in a workplace, it is almost always appropriate to consider the application of the *Criminal Code*. For an employer, not reacting appropriately to violence or sexual assault can have serious consequences. In some cases, the employer or its agent (for example, a supervisor) can be charged as a party to an offence. If the victim later launches a lawsuit based on the incident, the employer may in some cases be found vicariously liable (responsible for the actions of the violent employee) for the victim's injuries.

Smith (2000) suggests that violence can sometimes be predicted, especially where employees have a history, before the actual violent act, of extreme irritability or rage. Smith defines "rage" as loss of physical, verbal, or emotional control in a confrontation with another, and warns that employer tolerance of incidents of rage at the workplace can lay the foundation for violence. Smith, who has considerable experience as a trauma response consultant, cites studies that correlate the following employee characteristics with an increased risk for workplace violence:

- male gender;

- mid-thirties to mid-forties;

- white Anglo-Saxon;
- aggressive personality with a tendency to blame others;
- rigid value system, sometimes including religious fundamentalism;
- a recent life stressor, such as a job loss, demotion, and so on;
- an addictive personality and/or active addictions;
- a person who expresses paranoid thoughts;
- social isolation/being a loner;
- person with a history of confrontational behaviour;
- person with an unusual interest in weapons or the military;
- person suffering from a diagnosed or undiagnosed mental or emotional illness.

These factors, of course, do not mean that all white males in their forties will react to a demotion by bringing a gun to work. However, it may be useful for employers to know that the age range most closely correlated with workplace violence is perhaps higher than most people expect; or that loners, who generally avoid others, may in fact turn out to be quite confrontational when push turns to shove.

When an act of violence occurs in the workplace, employers should call the police, and provide immediate first aid if necessary. Witnesses should be encouraged to cooperate with investigators. In some cases, where the incident has been traumatic (or has the potential to have been traumatic) for onlookers, employers should consider the appropriateness of crisis counselling or debriefing for the employees involved.

FACTS AND FIGURES

Workplace Violence Against Women

A 2004 study conducted by the US Department of Labor discovered that, for female employees, the second leading cause of death at US workplaces in 2003 was homicide (murder) (UnitedStates Department of Labor, 2004).

It can be easy to dismiss this sobering statistic by assuming that workplace homicide is a US issue—Canada has a reputation as a much less violent society. However, the differences between American and Canadian violent crime rates relate primarily to crime *outside* the workplace. According to 1998 data collected by the Geneva-based International Labor Organization (ILO), the rate of physical assaults against women in the workplace is four times higher in Canada than in the US.

Smith cautions that not all victims and witnesses benefit from crisis or trauma counselling, and there is no one-size fits-all approach. Some employees may prefer to seek support from friends and relatives; some may not be ready to discuss the incident until considerable time has passed; some may find "debriefing" to be traumatic in itself (in that it may feel like they are reliving the experience); and yet others may be generally unaffected and may find forced counselling to be an invasion of privacy. The best approach to the aftermath of violence in the workplace may be to arrange for the availability of counselling and to inform employees about how to access it, without requiring anyone to use it. For employees who *do* decide to seek counselling, it is important that it be provided free of charge and, where possible, at the workplace during working hours, without deduction in pay for attending (because, after all, the triggering event will have occurred under similar circumstances—that is, at work).

Some employers have employee assistance programs (EAPs) in place already, with counselling being one of the services offered. Where this is the case, it is worth determining whether the EAP provider has appropriate trauma-response expertise and whether they can provide additional or special programs to address the specific event. Employees should be reminded of the program's existence and of the appropriate contact information. A useful way to do this is to arrange for EAP representatives to deliver a brief presentation at the office, introducing themselves and their services. Having names and faces to associate with this kind of benefit can help overcome any "fear of the unknown" that may dissuade employees from seeking out these services. EAPs will be discussed again in chapter 8 of this book.

Employers should realize that, despite their best efforts to react appropriately to workplace violence, it is possible that victims or witnesses will suffer post-traumatic-stress-related absences or problems in the aftermath of such an incident. The case law decided under the WSIA does recognize certain kinds of post-traumatic stress as valid grounds for a claim.

Criminal Negligence

Negligence, in law, is generally defined as the culpable (at-fault) failure to discharge a duty of care to another person. This can mean failure to take action where action is required, failure to prevent harm where prevention is required, failure to carry out a duty to an acceptable standard, or failure to consider (and therefore avoid) the harmful consequences of one's actions.

In civil (non-criminal) courts, negligence is a well-recognized basis for lawsuits in tort. To prove negligence in a tort lawsuit, a plaintiff need only show that the defendant owed her a duty of care (for example, the duty that store owners owe shoppers to keep the shopping experience safe by cleaning up spills, and so on) and that the defendant failed to carry out that duty.

In the normal course, negligence cases are dealt with under the civil law. The reason for this is that these cases generally involve non-malicious acts. A shop owner will typically not *intentionally* spill shampoo on the floor to create a slipping risk. The law generally recognizes that negligence cases lack the *mens rea* (a guilty mind or intention) element that makes a criminal charge—and the threat of a criminal record, prison, and so on—appropriate.

In some cases, however, an act of negligence has potential consequences so serious that it does begin to share the characteristics of a crime. For example, leaving a child in a car with the windows closed on a hot day while shopping constitutes criminal negligence. Swerving a car into oncoming traffic in a no-passing zone out of impatience and causing an accident is probably also criminal negligence (whether it is labelled that or labelled "dangerous driving" has little relevance—the theoretical basis for criminal liability is the same).

As you learned in the account of the Westray mine disaster in the Hard Lessons feature in chapter 2, recent changes to the *Criminal Code* have made it easier for police to lay criminal charges against employers in the wake of acts of serious negligence.

A new provision also makes it clear that "everyone" who directs the work of another has a legal duty to take reasonable steps to protect that person from bodily harm (see *Criminal Code*, sections 22.1, 22.2, and 217.1). While the words "criminal negligence" are never mentioned in this provision, criminalizing what would otherwise be *civil* negligence is what the provisions actually do. The sections are broadly and generally worded; in the context of the workplace, it could apply to any kind of "everyone" and any kind of task. These provisions put employers on notice that serious acts of negligence that might once have led to WSIA claims, OHSA prosecutions, and—in rare cases—civil lawsuits can now also form the basis for criminal charges.

FACTS AND FIGURES

Corporate Versus Individual Officer Liability— What's the Difference?

Before the recognition by the *Criminal Code* that a corporation *itself* can be a party to an offence, it was possible to charge individuals within the corporation with crimes, in appropriate cases. Given the fact that negligence is a human failure, if managers and corporate directors already existed as potential targets of criminal liability, why did the legislature feel it was necessary to extend this liability to the corporation proper? And what's the difference?

The answer is twofold. First, it can be very difficult, from a practical evidentiary perspective, to pinpoint a specific individual who is at fault in the wake of an accident. Second, some accidents are the result of the actions of more than one person. It is possible to imagine a situation in which, while none of the actions of each individual are negligent, the combined effect of the actions adds up to negligence. Consider this example:

> A forest fire is raging in a provincial park. A team of park staff is working to soak the forest floor in places not yet affected by the fire. The ground-level team is protected by a team of sentries posted in watchtowers above the treeline. The job of these sentries is to report on any sudden changes in the wind that might lead to movements of the fire that would endanger the ground crew. The park has no specific policies in place to schedule bathroom breaks for the sentries, who must climb down from the towers to privies on the ground. One day, all six sentries happen to climb down for the washroom at the same time. The wind direction shifts suddenly, and a ground-level worker is injured when a flaming timber falls on him as the fire approaches unexpectedly.

In this scenario, no individual sentry will likely be found negligent; however, the park itself (a corporation) may be negligent through failing to establish a policy governing coverage for washroom breaks.

KEY TERMS

technical standard

legislated technical standard

voluntary technical standard

REVIEW QUESTIONS

1. Why doesn't the Industrial Establishments regulation prescribe standards for the dimensions and construction of mine shafts?

2. List at least three circumstances, at a workplace, that would likely trigger the need for a Pre-Start Health and Safety Review.

3. What is the legal/compliance significance of the practice, by legislative drafters, of "adopting" technical standards developed in the private sector into regulations?

4. Which legal framework regulates safety during the transportation of dangerous goods in Canada? Which system regulates their use within a workplace?

5. What consequences could befall an employer who allows an uncertified worker to repair the brakes on a passenger bus (motor coach)?

6. Why, if standards like CSA-Z1000, Occupational Health and Safety Management don't have the force of law, might it be wise for an employer to choose to comply?

7. Can an employer discriminate against a disabled job applicant on the grounds of safety concerns?

8. An employer who has had problems in the past with drug-addicted workers advertises a set-painting job at a theatre. To avoid further disruptions to the work schedule by the need to fire anybody later, can she require that applicants submit to a urine-sample-based drug screening test as part of the hiring process? Is your answer any different if the test is administered after a conditional offer of employment is made?

9. Why should employers be alert to, and use discipline to deter, acts of verbal abuse or rage in the workplace even where these incidents result in no harm to workers?

CHAPTER 7

OHS from a Strategic Perspective

CHAPTER OBJECTIVES

After completing this chapter, you should be able to:

- Explain why some analysts believe that occupational health and safety is not an appropriate subject for a cost–benefit analysis.

- List at least three potential costs of workplace accidents other than WSIA premiums.

- Explain how a worker's absence affects productivity.

- List at least five different types of costs associated with occupational health and safety compliance.

- Describe at least two potential sources of funding to help offset occupational health and safety compliance costs.

- Explain how an employer might attempt to measure the effectiveness of its occupational health and safety program.

INTRODUCTION

In a presentation he gave at a 2006 Canadian Society of Safety Engineering (CSSE) conference, safety professional and CSSE Education Director Peter Sturm touched on how workplace safety professionals can best combat their image as "necessary evils" in the workplace.

This negative view of safety specialists is most often voiced in the stream of business thinking that lumps business functions into two main categories: "things that earn us money, and things that don't." The popular term for a business function that falls into the "things that don't" category is "cost centre."

In reality, there are many key professionals in modern corporations who cannot draw a direct connection between what they do at work all day and a chunk of the bottom line. Consider, for example, the work done by IT specialists whose function is to protect the confidentiality of sensitive business information, or the routine tests performed by quality control personnel who are hired to detect defects in a company's products before they reach the marketplace. Almost 100 percent of the time, the testing done by these individuals reveals that everything is fine. But what about the 1 or 2 percent of the time that it's not?

As this section was being written, the news in Toronto was dominated by reports of family pets across North America becoming sick and dying after consuming pet food produced by Menu Foods, a Streetsville, Ontario company. Preliminary testing suggested that the food might have been tainted by a rat poison that is banned in Canada but that may have made it into animal food via an ingredient imported from China. At the time of writing, the cost of the recall announced by the companies involved was estimated to exceed $40 billion, and Menu Foods had just announced its commitment to pay the vet bills for the treatment or euthanasia of hundreds of pets. **Class action** lawsuits were being launched in both the United States and Canada. Perhaps the most significant cost of all, however, will ultimately be to the company's reputation in the marketplace. If Menu Foods had had in place a program that would have identified the problem earlier, there would be no question, today, of the connection of that "cost centre" business function to the company's profitability.

The Menu Foods public relations fiasco was related to product safety, but consumer confidence can also be shaken in the wake of accidents that hurt the workers who produce the goods we use. Consider, for example, the work of agencies such as Canada's Ethical Trading Action Group (ETAG), the mandate of which, as stated at en.maquilasolidarity.org/about/etag, is to advocate "for government policies, voluntary codes of conduct, and purchasing policies that promote humane labour practices based on accepted international labour standards." See also the Profile feature below about the Maquila Solidarity Network's "toy campaign." While these groups and others like them tend to focus primarily on serious worker safety abuses overseas, catastrophic accidents happen in Canada, too—as you learned from the summary of Nova Scotia's Westray mine disaster in chapter 2. That accident led not only to the complete abandonment of the mine (at a cost of millions of dollars), but also to the passage of legislation permitting criminal charges against corporations in the wake of serious accidents.

class action
a legal action brought on behalf of the members of a group with a common grievance or interest

PROFILE

Maquila Solidarity Network's Toy Campaign

The Maquila Solidarity Network is a watchdog agency dedicated to monitoring safety, human rights, and other worker abuses in manufacturing, particularly in Mexico, Central America, and Asia. The agency is based in Toronto, Ontario and is part of the broader Ethical Trading Action Group (ETAG).

The Network describes its initiatives—which generally include invitations to question or boycott retailers selling items produced in substandard factories—as "campaigns." The agency's "toy campaign" was launched in response to workplace accidents in toy manufacturing companies, including:

- a May 1993 factory fire—possibly the world's worst ever—at Kadar Industrial Toy Company in Thailand. The fire killed 188 workers and injured 500 more due to lack of suitable fire exits; and

- a November 1993 fire at the Zhili Handicraft Factory in Shenzhen, China that killed 87 and injured 46 workers.

> Another motivator for the campaign was the existence of poor working conditions in much of the Asian toy manufacturing industry, which employs very young workers and imposes very long working hours at a very low rate of pay (often about 25¢ per hour).
> The Maquila Solidarity Network's campaign involved advising toy consumers not to boycott the industry but rather to urge North American toy retailers to insist that suppliers sign a safety "code of conduct." By doing so, suppliers agree, among other things, to allow factories to be monitored by local (that is, in the source countries) human rights monitors.
> Other Maquila Solidarity Network campaigns have targeted "household name" manufacturers and retailers using strategies such as the awarding of a "Sweatshop Retailer of the Year" title after inviting online voting by concerned consumers.

Finally, because safety is often viewed as a moral or ethical issue, neglecting it can have an impact on investment in the company by would-be shareholders. There has been an increase, in recent years, in the popularity of ethical investment—investment in companies that demonstrate superior ethics in a variety of contexts, for example, environmental responsibility or worker safety. Investors seeking to make only "ethical" investments may prefer companies with a strong safety track record. Consider this official statement by asset management and investment company Calvert Group, Ltd. (2008):

> Calvert believes that companies with strong workplace policies and comprehensive labor and safety programs tend to perform better than their counterparts. Conversely, Calvert believes that unsafe work environments, unfair labor practices, and poor wages and benefits have negative effects on a company's ability to attract and retain talented employees, while also damaging the company's public image and reputation.
> Therefore, Calvert prefers companies that comply with federal and state regulations regarding wages, hours, child labor standards, and other conditions of work. Calvert avoids companies with a pattern of violating fair labor practices and occupational safety and health regulations. Calvert also avoids companies that are subject to serious action by US regulatory agencies such as the Occupational Safety and Health Administration.

As you can see, any perspective on workplace health and safety that fails to consider the relationship between worker safety initiatives and a company's financial health can probably be dismissed as short sighted. This is true even where only *potential* harm to the company's public image is concerned. As you will learn in this chapter, however, worker injuries also influence the cost of production, and devoting appropriate resources to preventing them makes good business sense. Before you read this chapter's business case for OHS, however, it is useful to take a step back and consider the question posted in the following section.

SHOULD OCCUPATIONAL HEALTH AND SAFETY BE SUBJECT TO A COST–BENEFIT ANALYSIS?

While this chapter will attempt to demonstrate a solid business case for the promotion of worker health and safety, many people believe that the very exercise of considering whether or not one can afford to run a safe shop is morally repugnant.

Should a business owner be permitted to weigh the potential cost of workers' lives or impaired health against the cost of technologies designed to protect those workers, and act on a finding that it is probably cheaper to put lives at risk?

A teaching document prepared for Minerva Technologies by Oshawa's University of Ontario Institute of Technology (UOIT) describes a scenario in which an engineering firm, faced with indications that a job is not safe, must choose either to absorb the cost of a production delay or press on after characterizing the risks to workers as a cost of doing business. The authors of the teaching document characterize choices like these as being fundamentally different from other business decisions, because they have an *ethical* component (Rosen, 2004).

Workers and the unions that represent them are, understandably, particularly critical of the use of cost–benefit analyses to guide OHS decisions. The notion that a workplace is as safe as the employer can afford to make it—as opposed to just plain *safe*—is objectionable to most workers, and can motivate decisions to work elsewhere. As well, the *Criminal Code* of Canada now makes it clear that it is the employer's legal duty to take "reasonable steps" to protect the safety of workers. Taking too narrow a view of what is "reasonable" exposes employers to criminal justice consequences.

Finally, the very nature of safety initiatives makes it difficult to accurately measure the positive consequences of those initiatives. Most safety efforts are designed to prevent things—injuries and illnesses—from happening, which makes it difficult to assess progress: after all, how do you count the number of accidents that *didn't* happen? The challenge posed by attempts to measure safety "results" means that we may often underestimate those results, and comparing our estimates against the concrete figures for compliance costs does *not* amount to an analysis reliable enough for the purpose of making business decisions.

In his book *Working Safe: How to Help People Actively Care for Health and Safety*, Geller states that it is unfair to impose a cost–benefit analysis on a function whose goals are not subject to measure:

> [T]he primary purpose of getting involved in safety intervention is to prevent injury.... Unfortunately, we rarely see these most important consequences. Thus, we need motivation: feedback, interpersonal approval, and self-talk. We tell ourselves that the safe behavior is "the right thing to do," and that someday an injury will be prevented. We can't count the number of injuries we prevent; we just need to keep the faith. (Geller, 1996, p. 11)

Finally, as noted in the introduction to this chapter, there are reasons to believe that a commitment to safety may have a significant impact on profitability even where it is not possible to measure that impact as it relates to an individual company using traditional cost–benefit analysis. In 2003 a team of researchers (Orlitzky, Schmidt, and Rynes, from the United States and Australia) conducted a "meta-analysis" of the results of 52 different studies that had attempted to capture the relationship (if any) between corporate social/environmental performance (CSP, which includes performance in protecting worker safety) and corporate financial performance (CFP). The mixed results of these individual studies had led to a general view, in the business world, that there was no clear connection between being socially responsible and being profitable. However, the results of the Orlitzky, Schmidt, and Rynes meta-analysis challenge that view, and suggest that "social responsibility and, to a lesser extent, environmental responsibility is likely to pay off." Taking account of all of the relevant studies considered together, "[t]his meta-analysis establishes a greater degree of certainty with respect to the CSP–CFP relationship than is currently assumed to exist by many business scholars." In other words, a corporation's efforts to be socially responsible—while often not directly measurable by means of a straight cost–benefit analysis—are more likely to improve rather than hurt the company's profitability.

THE COST OF WORKPLACE INJURIES

Introduction

In chapter 5, you were introduced to the Workplace Safety and Insurance Act (WSIA) and learned that an insurance-based system has now replaced private-sector lawsuits for workplace injuries. While there is some variation in premiums associated with performance, in general, the WSIA system provides predictability for employers when it comes to the cost of compensating injured workers. However, this predictability relates only to the costs of the compensation claimed under the WSIB system (lost time, health care, future economic loss, and so on).

It is important to realize that the costs covered by the WSIB fund represent only a fraction of the total estimated costs of a workplace injury. In fact, most analysts agree that the total costs of an average workplace injury are at least *four times* the amount paid out in compensation to the injured employee.

Business Results Through Health and Safety, a 2001 publication developed by the CME and WSIB, reports that while the average lost-time injury claim filed with the WSIB in 1999 was over $11,771, the *total* estimated costs to the employer of an "average" injury were closer to $59,000. How can this be?

The answer is that when a person is hurt at work, there are a wide range of different kinds of costs, many of which are not reflected in WSIA premiums.

WSIA premiums and the incentive programs that influence them (NEER, MAP, CAD-7) were discussed in chapter 5, and we will not revisit them here. The following sections describe the employer's injury-related costs that go above and beyond WSIA premiums.

There are tools available to employers to assist in the calculation of non-premium incident/accident costs. One example is the Costing an Individual Incident/Injury form provided by the WSIB as part of its *Business Results Through Health and Safety* document (a simpler worksheet for the same purpose is included in the WSIB's *MAP Cost Tool* document, which is intended for small business). This form, which is included in appendix F, encourages employers to take into account easily overlooked injury costs, many of which will be mentioned below. The Ontario Safety Association for Community and Healthcare (OSACH) offers a similar tool for estimating the costs of a workplace injury or illness in the health care sector (see www.osach.ca).

Property Damage

Many workplace accidents lead to damage to the physical plant or equipment, or to product. There may be costs related to cleaning up spills or debris (for example, costs of cleaning products/absorbents or degreasing products, and costs of labour to perform the cleanup). Fire damage can be especially expensive to clean up.

Where machinery is damaged, the company should take into consideration not only the cost of repairs to the machinery, but also the cost of renting replacements while the repairs are made. In some cases, even where there is no damage to equipment, an investigation by the MOL may lead to orders prescribing the replacement or retrofitting of equipment implicated in the accident.

Finally, the cost of any spoiled product must be taken into consideration. Stopping a manufacturing process may lead to wastage, and product may be damaged through being spilled, dropped, or otherwise mishandled.

Lost Productivity

There are two kinds of personnel-related costs that can result from an accident: those that relate to the temporary or permanent replacement of the injured employee(s) (these will be discussed in the next section), and those that flow from changes to the work responsibilities of employees who remain on the job.

For example, if an accident causes debris or damage to the premises or equipment, production may need to be halted until the problems can be resolved. Some workers who normally work in profit-generating roles may instead need to devote their time to repairs or cleanup. Once production resumes, the remaining workers may need to work overtime to make up for delays, and supervisors will bear the burden of having to schedule overtime, arrange for replacement workers, and train replacements to do the absent workers' jobs.

After especially serious accidents, workers who witnessed the accident may suffer from lowered morale, which can influence productivity. Time off for counselling may be appropriate to assist some workers in dealing with the fallout from the accident, and this time away may require coverage by other workers, sometimes at a premium rate.

Recruitment and Training Costs

Where the injured employee is expected to be off work for an extended period or permanently, the employer will need to arrange for a replacement. The recruitment process, and the training that will follow it, both have an impact on productivity and cost money.

The impact of the loss of an injured employee is magnified in cases where the employee possessed unique skills. While the company searches for a replacement, work may be inefficient as other employees scramble to do his specialized job. Replacing him may take a great deal of time, or else the employer may need to pay for special training, education, and/or certification for the replacement worker. In some cases when an employee is off the job, the greatest loss is in the form of a loss of business advantage. Some employees are valuable to the company because of their broad knowledge of the business's operations, their experience in the industry, or their contacts. Such employees are irreplaceable, and the cost to the company when they are lost to injury—while difficult to quantify—is substantial.

If an injured employee is away for an extended period before she returns, or if she returns with a disability that requires accommodation, the employer will incur costs associated with retraining her to adjust to an altered—or totally different—job.

Administrative Costs

Accidents also entail considerable administrative costs. The accident must be documented, and sometimes investigated. This can require a time expenditure by the human resources department, as well as the safety engineering/occupational hygiene department (if there is one). Finally, senior management will need to review the results of any investigation and spend time coming up with suitable remedial actions—an effort that can sometimes involve the hiring of a specialized consultant.

As you learned in chapter 3, certain accidents must be reported to the MOL within fairly tight deadlines; should the MOL decide to conduct its own inspection or investigation, the employer will need to designate one or more workers (usually, the Health and Safety Representative or a member of the JHSC) to accompany MOL representatives during the inspection.

Payroll staff will need to complete paperwork to document the injured employee's absence from regular duty. The human resources department will also need to prepare the appropriate documentation to assist the employee in pursuing a claim under the WSIA.

Finally, where an employee is off work due to injury, the employer will be required to follow up with him on a regular basis to monitor his plans for return, to support him in preparing to return, and to assess the ongoing validity of his WSIA claim and his eligibility for continued health services.

In their discussion of the administrative costs of a workplace accident, the authors of *Business Results Through Health and Safety* raise the issue of opportunity costs. Opportunity costs are the difficult-to-estimate costs associated with lost chances. To understand opportunity costs, you can ask yourself the question "if only." For example, "if only Manager Brenda did not spend half of Thursday and all of Friday arranging for the rental of a replacement widget snipper and arranging for repairs to the existing one (the one broken when the injured employee fell on it), what would she have done with her time?" Chances are, had there been no accident, she would have been doing other work that contributes directly to the company's profits.

A company may also incur costs while working to prepare for safety inspections or audits. When a company's poor safety performance attracts the MOL's attention, the company may be referred to a program called Workwell, which conducts compulsory occupational health and safety audits. When a company fails an initial Workwell audit, it is given six months to make recommended changes. Making these changes can involve purchasing equipment, upgrading existing equipment, and hiring occupational health and safety consultants. The cost of making post-audit changes can easily exceed $10,000 for a small company, but failure to pass the second, follow-up audit can result in a penalty in the form of a percentage of the company's WSIA premium being imposed. The maximum penalty available is 75 percent of the premium, to a maximum of $500,000.

Fines and Other Penalties

In cases in which an MOL investigation of an accident reveals non-compliance with the OHSA, the employer may face fines. As noted in chapter 3, current maximum fines for non-compliance are $25,000 for individuals and $500,000 for corporations. A victim fine surcharge can be added to these fines to reflect the aggravating factor of having actually caused harm (as opposed to non-compliance discovered, for example, in the course of a routine investigation). Victim fine surcharges were discussed in chapter 6.

FACTS AND FIGURES
OHSA Fines

According to the Ontario MOL, in 2002 over $5 million in fines was levied by Ontario courts against employers for violations of the OHSA.

Other potential sources of monetary fines and penalties, in serious cases at least, are criminal charges and civil lawsuits. As you have read in previous chapters, recent amendments to the *Criminal Code* of Canada make it possible for a corporation to face criminal charges in the wake of a workplace accident characterized by criminal negligence. The maximum penalties available for accidents for which a corporation is convicted criminally will depend on the nature of the charges. In general, the penalties for crimes that cause bodily harm or death are expressed in the form of maximum (and sometimes minimum) prison sentences; however, where a corporation (as opposed to an individual officer) is charged, courts will need to impose fines in lieu of imprisonment. The fines that can be charged are largely at the court's discretion. For **summary conviction** offences, the maximum fine chargeable against an organization is $100,000 (*Criminal Code*, section 735). However, many of the offences most likely to be charged in the wake of an accident—for example, criminal negligence causing bodily harm, or criminal negligence causing death—are indictable (more serious) offences. This suggests that fines greater than $100,000 will be in order for these offences, where circumstances warrant.

As under the *Provincial Offences Act* (which applies to the OHSA), there is provision under the *Criminal Code* for the imposition of victim surcharges over and above the "basic" fine (*Criminal Code*, section 737).

Because the WSIA is designed to provide an alternative to civil actions to recover compensation for workplace injuries, there are significant limits on the circumstances in which a corporation will face a civil lawsuit flowing from a workplace injury. Companies that are not required to be covered under the WSIA (for example, due to very small size) are vulnerable to civil suits by injured workers. A company may also face a civil action where the accident involved victims who were not employees—for example, customers visiting a store during a roof collapse, or passersby injured during a factory explosion. Civil damage awards reflect the degree of the company's responsibility for the accident and the extent of the victims' injuries, and so cannot be predicted or prepared for easily (though some companies purchase private insurance to cover these costs).

Finally, while not exactly a "fine" or "penalty," a MOL order to repair or replace equipment may result from the investigation of an incident. Compliance with these orders is compulsory, and can be expensive. In most cases, the employer is prohibited from using the equipment or portion of the physical plant in question until the prescribed changes have been made.

summary conviction offence
a minor offence that is tried using simplified criminal procedures

Impact on Public Image

The potential impact of workplace accidents on a business's image in the marketplace was touched on in the introduction to this chapter. Principled consumers may hesitate to purchase goods or services from companies that they perceive as dangerous places to work.

From an investor perspective, it can be personally distasteful to feel that one's money is supporting an employer who is indifferent to the safety of its workers. But a poor safety record may have other, subtler effects on the attractiveness of a company to potential investors. Poor performance—demonstrated by failures to protect worker safety—may be interpreted by business analysts as evidence of mismanagement in general. Why didn't/couldn't the business identify the risks that led to a serious accident or predict the consequences? Do managers lack awareness of workplace conditions? Are supervisors ineffectual? Does the company employ incompetent labour? Is there a morale problem (either one that caused the accident, or one that flows from it)?

Costs to Employees, Families, and the Community

It is also worthwhile, from an ethical standpoint, to touch on the costs of workplace injuries to workers, their families, and the community. The $59,000 estimate of the injury costs for an average injury quoted above measured only those costs borne by the employer, not those borne by employees, their families, or the community.

A 1995 report by the Australian Industry Commission found the following:

> Past research indicates that injured employees bear about 30 percent of the total costs of workplace injury and illness, which include loss of income, pain and suffering, loss of future earnings, medical costs and travel costs. The share of costs borne by injured employees rises sharply with the severity of the outcome while the share borne by their employer falls. (p. 18)

Australia has a worker's compensation system quite similar to Ontario's.

The WSIA scheme is designed to ease the financial impact of workplace injuries on individuals, but it does not eliminate that impact entirely. Examples of the types of non-reimbursed costs that individuals may incur after an injury can include:

- costs related to injuries not currently recognized as such by the system (for example, harm in the form of stress or pain);

- costs related to the inability of some disabled individuals to reach their full potential (consider, for example, a young woman who is completely disabled while working at a summer job in an automated bakery in between university terms while she is studying to become an scientist for missions in outer space—she will receive WSIA benefits designed to partly replace her income as a bakery worker, not as an astronaut);

- costs related to harm to interpersonal relationships (for example, the strain on a marriage from an injured man's post-accident depression may lead to a financially disastrous divorce); and

- intangible costs related to the loss of activities that contributed to the injured worker's quality of life (for example, the loss of the ability to play a sport or musical instrument).

Costs that may be borne by the family or community include:

- costs related to a spouse cutting back his working hours to care for a wife who was injured at work;
- costs related to having to hire domestic help to handle tasks one used to be able to complete (for example, housework or child care);
- the loss, by one spouse, of the ability to pursue a joint hobby (for example, ballroom dancing or ecotourism); and
- the loss to the community that results from the injured worker's reduced ability to participate in activities that benefit the community (for example, volunteer work, political activism, performance in community theatre, and so on).

Socially conscious employers will be motivated to prevent injuries because they realize that the impact of those injuries has a "ripple effect" that extends beyond the workplace and reaches into the personal lives, families, and communities of those who have dedicated their efforts to the company's success.

HARD LESSONS
Managing the Special Vulnerability of Younger Workers

To employers, a young worker—often defined as a person under the age of 25—makes a very attractive employee. Young workers may be newly out of school, which makes their skills the freshest and most current available; they may have a great deal of ambition and energy; and they are often free of the established habits and attitudes that make it difficult to adjust to new tasks. Finally, their limited experience typically means that they are tolerant of lower wages than their more experienced counterparts would expect.

But as workplace injury statistics have proven in the past four decades, younger workers also come with a disadvantage: they seem to be more vulnerable to workplace injury and death than their older counterparts. Workers under the age of 25 form the age group at the highest risk of injury. There are multiple reasons for this special vulnerability, for example:

- Young workers, eager to make a positive impression, may readily take on risky tasks or tasks for which they lack training.
- Young workers may lack the self-assurance and assertiveness needed to ask questions about the safety of a task, the availability of protective gear, or appropriate procedures. They may also be less willing than their older colleagues to challenge an employer's unsafe practices.
- Young workers may accept unreasonable risks in order to keep a job because they view themselves as at a disadvantage, in the labour marketplace, compared with more experienced workers.

- Because young people often work in low-paying jobs for short terms, employers may be reluctant to invest training time and money in teaching job skills. A study by the Industrial Accident Prevention Association revealed that 56 percent of young workers reported having received no training before taking on a new work task.

- Even where young workers *do* get training, the training they receive may be task-related "splinter training" (in the sense of learning just a splinter of the normal training program). The WSIB's study *Managing the Health and Safety Interests of Young Workers in Small Business* noted that young workers lacked *general* OHS training, for example, training about the need to think about how to manage risks and their own safety in the workplace. (Blanco & Lewko et al., 2001)

To make matters worse, not only are young workers at an increased risk of injury when compared with older workers, their injuries cost more to employers and to society. The reason for this is that where a young worker is permanently disabled by injury or illness (even partly disabled), the loss of earning potential (and the loss of opportunity to contribute to society) is higher when the victim is younger for the simple reason that there are more years of working life ahead of a young worker. This sense of loss is reflected in the public's tendency to react with sadness and outrage to news that a young person was hurt or killed on the job.

The statistics about the vulnerability of young workers need not frighten employers away from hiring them, because, as you can see from the list of factors that contribute to young workers' injuries, many of these injuries can be prevented by encouraging young workers to participate fully in the IRS. When young workers are taught that they are not only *entitled* but *expected* to work safely, many of the behaviours that lead to injuries can be prevented.

COMPLIANCE AND PREVENTION COSTS

Categories of Costs

The costs of promoting safety at work fall into a number of general categories, including:

- WSIA premiums, including supplementary premiums under incentive programs;
- top-ups provided to injured employees (if offered);
- costs associated with health care services offered at the workplace (for example, first aid or trauma counselling);
- fines and penalties levied under statutes or under the criminal or civil law;
- the cost of personal protective equipment (PPE);
- the cost of protective mechanisms on equipment or in the physical plant;
- costs related to upgrading equipment or facilities, either as a function of regular maintenance/modernization or in compliance with government orders;
- administrative costs related to documenting claims, conducting investigations, supporting the work of a JHSC, and managing ongoing claims and worker absences;

- costs related to health and safety benefit programs, for example, EAPs;
- safety training and certification costs, either in-house or from outside providers;
- costs of safety promotion campaigns (signage, posters, program developer/delivery time); and
- the cost of incentive programs, for example, awards and prizes.

Justifying OHS Costs

Some business owners worry about how to justify the costs of health and safety compliance to their shareholders. However, when viewed as a risk management measure, these costs become more palatable. The authors of *Business Results Through Health and Safety* note that preventing injuries amounts, in many cases, to ensuring against the risk of business interruptions:

> Ignoring or underplaying risks that may result in major interruption of the business is simply unacceptable from a shareholder or regulatory perspective. … If an incident is serious (fire, explosion, significant employee illness due to toxic exposure etc.), it can put the operation out of business. This is not bad luck. It is bad and negligent management. (CME and WSIB, 2001)

When this perspective is considered, it becomes clear that attributing safety-promotion-related costs exclusively to "compliance" is artificial. In many cases, injuries and illnesses occur because of problems in the business that can affect productivity in other ways. For example, if an employee is injured while recalibrating a piece of machinery, an investigation may reveal that the equipment required frequent adjustment and that its repeated malfunctions were causing unacceptable wastage of raw materials. Or if an employee was sickened because she was not properly trained about how to handle hazardous chemicals, the employer may learn, through the accident, that its safety and environmental policies are not being communicated by supervisors to front-line staff. Is a costly environmental accident at risk of happening because of mismanagement? Safety-related investigations often reveal root causes that have important productivity-related implications. Making a workplace safe is *integral* to good business:

> Most of the businesses consistently achieving lower injury/illness rates have done this by using sound management practices. They have made safe operation a business objective, not just because of concern for their employees, but because safe operation is a factor in their continuing business profitability and success. (CME and WSIB, 2001, p. 2)

Thinking Safe

Workplace Health Promotion Programs

In a February 2004 article in *Newsbeat*, a newsletter of the Canadian Association of Cardiac Rehabilitation (CACR), Linda Makrides, PhD, then Director of the Atlantic Health and Wellness Institute, explored the growing trend among workplace safety leaders to incorporate a workplace health promotion (WHP) program into their existing OHS program. WHP programs are primarily education-based programs designed to promote employee health by raising awareness of health risks such as smoking, substance abuse, sedentary lifestyles, and poor diet.

The rationale behind these programs is that the workplace is a convenient setting in which health promotion information can be communicated, leading to potential savings for the employer (in a decrease in disability claims, sick time, and so on). Makrides quotes a 2001 World Health Organization (WHO) health report as follows:

> [The workplace is] the single most important channel to systematically reach the adult population through health information and health promotion programs. (Makrides, 2004, p. 2)

WHP programs can be tailored to fit the needs and the resources of the workplace—they can be as simple or as elaborate as seems fitting for the setting. Simple programs involve encouraging participants to use self-assessment tools, such as questionnaires, and to read educational materials. More elaborate programs may employ focus groups and incorporate a lifestyle coaching/counselling service. An important aspect of all successful programs, however, is that health promotion/illness and injury prevention activities are integrated into the process of workplace change—change, for example, in work processes. Makrides cites studies that demonstrate that good WHP programs effectively lower health care and insurance costs, reduce absenteeism, and improve performance and productivity. She reports on cost–benefit analyses conducted by employers who have used these programs, finding an average return on investment (in terms of cost savings) of approximately $2 to $6 saved per every $1 invested, with some employers reporting returns of over $10 saved for every $1 invested.

Opportunities to Offset OHS Costs

INTRODUCTION

To encourage employers to commit to safety, a number of programs exist that can help offset the costs of safety compliance. The NEER, CAD-7, and MAP incentive programs—which offer WSIA premium rebates for strong safety performers—were discussed in chapter 5; the Second Injury Enhancement Fund (SIEF), a special WSIB program that seeks to offset some of the risks of hiring previously injured or disabled workers, will be highlighted in chapter 11. Still other programs are available for employers willing to put in the time to investigate them.

SAFETY GROUPS

The Safety Groups Program is a WSIB-sponsored program designed to facilitate the business-to-business transfer of safety knowledge and best practices. Schedule 1 employers who are in good standing with the WSIB are eligible to seek membership in an appropriate safety group.

There are approximately 30 safety groups in existence, with the potential for more to be developed should interest allow. Safety groups can be industry-specific (for example, there are several "construction" groups, a few "automotive" groups, and one "education" group) or general in focus but regionally based (like the Kingston Partners for a Safe Community).

Each safety group has a "sponsor"—a member organization that serves as the group's liaison with the WSIB. For example, the education group's current sponsor is the Education Safety Association of Ontario.

Each year, member firms commit to initiating at least five actions from an Achievement List prepared by the WSIB to guide the work of the group. If a member meets the program's participation requirements (which include performing a baseline OHS performance analysis and attending at least three meetings annually), it becomes eligible for a premium rebate. Rebates are determined based on the performance of the group as a whole: even if an individual member meets 100 percent of the participation requirements, the rebate the group receives will be based on the average performance for the group.

SAFE COMMUNITIES INCENTIVE PROGRAM

The Safe Communities Incentive Program (SCIP) is designed for small business, and provides a fixed premium rebate of 5 percent upon successful completion. Firms that participate in SCIP are not eligible to participate in the Safety Groups Program, which is targeted primarily at larger organizations, but SCIP participation does not preclude the receipt of additional rebates through the NEER, MAP, or CAD-7 programs.

A business is eligible if it pays less than $90,2000 in annual WSIB premiums and is located in a participating community. A list of participating communities is available on the WSIB's website.

When a business signs up for SCIP, its owner or a senior manager must complete a 16-hour program that covers the following topics:

- Leadership in health and safety
- The internal responsibility system
- Hazard identification and assessment
- Hazard controls and emergency response
- Return-to-work and claims management

At the end of the program, the participant must submit a health and safety policy and a statement of health and safety goals to the WSIB for review. Upon the WSIB's approval, the 5 percent premium rebate will come into force.

MEASURING RESULTS

As you have seen, it can be very difficult to measure the results of occupational health and safety initiatives.

The most readily available tool for measuring safety are the company's NEER, CAD-7, or MAP statements. These incentive programs adjust premiums based on the company's safety performance relative to the performance of other similar businesses. The WSIB reports that top safety performers can out-perform the bottom 25 percent (quartile) of businesses by 70 to 90 percent. Receiving a premium reduction under an incentive program is a sure sign that the company's efforts are effectively reducing workplace injuries and illnesses.

Another way in which companies can measure results is by keeping accurate records of injuries and incidents of work-related illness. Many of these incidents need to be reported to the WSIB, and so records will be generated in the course of the reporting process. However, it can be useful for a company to supplement the basic compliance-level record keeping with internal analysis that looks deeper—into the root causes of the accident. By taking the time to think about the causes of an accident (either in the course of a full internal investigation or more informally, for less serious incidents) employers will be better equipped to avoid future problems. From a cost perspective, applying the analysis developed by the WSIB (reproduced in appendix F can assist the employer in building a strong business case for safety improvements.

A precise cost–benefit analysis of a specific safety measure is rarely possible. However, there are credible statistics that suggest that spending money to improve safety is a good investment. In its *Fall Facts* brochure, Canadian fall protection company Innovative Fall Protection cites the US Department of Labor's estimate of a cost savings of between $4 and $6 for every $1 invested in good quality health and safety programs (Innovative Fall Protection Inc., 2003).

Earlier in this chapter, you read that the authors of *Business Results Through Health and Safety* estimated an employer's average per-injury costs at $59,000. The authors of that report also provided an insight into the potential savings to a company's productivity through the prevention of just one "average" lost-time injury:

> If your profit margin is 10%, it requires $590,000 in sales to produce $59,000 of profit. Alternatively, if your profit margin is 6% for example, it requires almost a million dollars ($983,333) in sales to produce $59,000 in profit. A reduction of a lost time injury costing $59,000 has the equivalent profit effect as increasing sales by $590,000 at a 10% profit margin or $983,333 at a 6% profit margin. (CME and WSIB, 2001, p. 8)

In a 2001 seminar, Ted Pattenden, former CEO of National Rubber Industries (NRI), reported measurable results following an effort to create a culture of safety at his company:

> Safety is a positive cultural element that leads to other business improvements. In 1990 NRI had 3800 lost days, in 1999, 16 lost days. In 1990 NRI was losing money, In 1999 NRI had top quartile profitability. A primary emphasis was changing the safety culture.
>
> Management that cannot manage safety cannot manage other functions. Symptoms may include:
>
> - Poor housekeeping
> - Many unscheduled equipment breakdowns
> - Lower employee morale and high turnover
> - High cost—failure to meet profit targets
>
> Employees need to know that they are important and that management cares about their welfare.

Whether or not your organization chooses to conduct a cost–benefit analysis of safety promotion expenditures, statistics make it clear that injury rates (and related accident costs) *can* be reduced by appropriate health and safety programs and through efforts to create a "culture of safety" within the workplace. You may even believe that you have a moral or ethical duty to create such a culture. Discharging that duty, while difficult in the short term, may well set your company apart from the pack, establishing it as a safety leader. Consider the following quote:

> The need for business survival challenges ethical conduct, and the choice of profit over ethical conduct tends to gamble with the probability of corporate demise.

Conversely, ethical choices may not guarantee or even lead to business success/survival, but if the company survives, such choices will most certainly lead to public confidence in the company or even the industry. (Patankar et al., 2005)

SAFETY LEADERSHIP

The use of the term "accidents" to describe workplace events that lead to illness or injury is not ideal, because it can leave the impression that these events are unpredictable, random, and uncontrollable. Often, this is not the case. While not every circumstance or combination of circumstances that can arise in a workplace can be foreseen, the fact that it is indeed possible to reduce accident rates by taking specific action has been proven over and over again.

If accidents were truly random, the ranking of companies based on safety performance would vary widely from year to year. In fact, it does not. There are clear "safety leaders" in most industries: employers that enjoy consistently better-than-average records with respect to worker safety.

> ### Thinking Safe
> #### Leading with Safety and Boosting Profitability
> Interestingly, as noted earlier in this chapter, companies that are recognized as safety leaders also tend to be leaders in terms of profitability. This correlation has not escaped the notice of experts who study management theory. In fact, one such expert—Thomas R. Krause, author of *Leading with Safety*—has suggested that CEOs hoping to achieve a performance-boosting culture shift within an organization might do well to choose safety promotion as the cornerstone of their culture change initiative. Krause identifies safety as a "metaphor for organizational excellence," and his research suggests that a demonstrated commitment to safety at the executive level can have a direct, positive influence on the organization's excellence in other areas, such as productivity.

When the safety performance of safety leaders is compared with that of "safety stragglers," the difference in safety performance is not slight—as one would expect it to be if accidents were random. Instead, it is fairly dramatic. The authors of *Business Results Through Health and Safety* report that, according to the WSIB, compensation and other accident-related costs for safety leaders (those at the top of performance rankings) tend to be 70 to 90 percent lower than those of the companies performing in the bottom quartile. Safety is not a matter of luck, but rather a result that flows from identifiable actions.

Good workplace injury and illness statistics may define a company as a safety leader, but they do not provide sufficient information for companies looking for an example to follow. What, specifically, are the factors that predict good safety performance?

One factor is a conscious choice, on the part of the company, to change its perception of safety. Instead of viewing workplace health and safety as a compliance issue—a "must-do" necessary evil—safety leaders view safety as an organizational core value, committing to it independent of the regulatory context.

Compliance with occupational health and safety laws is a minimum legal requirement. Maintaining minimum compliance will protect a business from fines and from excessive interference by regulatory bodies, but it will be insufficient to pay the dividends—in terms of public image, productivity, and good employee relations—that flow from being a safety leader. Safety leaders approach injury and illness prevention not as a compliance obligation (something that must be done to stay out of trouble) but rather as a goal worth pursuing, for the sake of workers and their families and for the impact safe operation can have on the business's performance.

HARD LESSONS
Is There No Such Thing as an "Accident"?

DuPont Canada, a company with a strong safety record in recent years, embraces worker safety as one of its core values. In a summary on its website (at www2.dupont.com/DuPont_Home/en_ca) of the company's safety beliefs, the following bold statement is made: "All injuries and illnesses are preventable."

This belief and its expression are fairly common in health and safety circles. In fact, Ontario's own WSIB recently launched an injury reduction advertising campaign entitled "Road to Zero." However, while workplace accidents and injuries have decreased dramatically in the period since about 1980, a zero accident rate continues to lie well beyond the reach of most large employers.

Some safety specialists are critical of approaches to safety that proclaim that there is no such thing as an accident. These critics argue that "zero tolerance" policies may encourage investigators to focus on assigning blame instead of focusing on developing prevention strategies designed to prevent a repeat incident.

Workplace accidents have *mostly* predictable causes, including:

- a lack of safety policies and procedures in the workplace;
- insufficient training/education about safety;
- insufficient effort put into identifying, investigating, and minimizing workplace hazards;
- insufficient effort put into adapting safety policies, procedures, and training when new processes or equipment are introduced; and
- failure on the part of workers, even where training and procedures are sufficient, to follow safety procedures.

Occasionally, however, unique circumstances cause an accident to happen in a way or for a reason that could not have been predicted easily, even in a workplace where safety is a priority. An example might be an accident, one of the causes of which is an unusual natural phenomenon, like a severe hailstorm.

Where an injury arises out of a "freak" accident, criticism of workers or supervisors based on a philosophy that "there are no accidents" can be unhelpful as well as potentially damaging to morale.

While it is important to investigate the root causes of unexpected accidents, it is important to make a distinction between finding *causes* and finding *fault*. An emphasis on "zero tolerance" for accidents combined with an investigative process that focuses on fault and a philosophy that "there are no accidents" can lead to a climate in which workers and supervisors hesitate to report "near misses," and may even conceal actual accidents.

KEY TERMS

class action

summary conviction

REVIEW QUESTIONS

1. What is a "cost centre"? Why has occupational health and safety sometimes been considered a cost centre?

2. What is "ethical" investing? Why might an investor committed to this principle prefer companies with a good safety record?

3. Why do some business analysts believe that it is immoral or unethical to apply a cost–benefit analysis to health and safety spending?

4. Do efforts, on the part of a company, to be socially responsible automatically eat into the company's profitability?

5. List at least three kinds of costs, other than the cost of WSIA premiums, that can flow from a workplace accident.

6. How do workplace accidents generate administrative costs?

7. What arguments might a business use to justify its OHS expenditures to shareholders?

8. List at least three sources of cost breaks available to employers who are motivated to improve safety performance.

9. What are the key differences between an organization that is simply compliant with occupational health and safety regulations and an organization that is a safety leader?

CHAPTER 8

Steps to Compliance I

CHAPTER OBJECTIVES

After completing this chapter, you should be able to:

- Describe the basic steps involved in developing or renewing your company's occupational health and safety program.

- Explain why the involvement of senior management in the program development/renewal process is essential to success.

- List the areas that should be assessed in the course of a baseline analysis of an organization's safety status.

- Understand the importance of establishing safety priorities.

- List the implementation tasks that commonly form part of the OHS renewal process.

- Identify human and learning resources that assist with the OHS development/renewal process.

- Explain the need for ongoing review, evaluation, and goal setting.

INTRODUCTION

The previous seven chapters have described and put into context the laws and regulations that touch on occupational health and safety, and have introduced the business and strategic reasons behind a commitment to working safe. This chapter will provide a road map—made up of specific tasks—that will help you prioritize and initiate the steps your organization needs to take in order to become a safety leader.

The authors of *Business Results Through Health and Safety* advocate breaking up the daunting job of improving occupational health and safety into manageable goals, while adopting a sense of urgency and setting aggressive deadlines for change:

> Plan your implementation in manageable chunks. Be aggressive, but do not over commit. It is far better to implement and enjoy success with a few items than to experience frustration in trying to make too much change at once. (CME and WSIB, 2001, p. 57)

The basic formula for managing any kind of change in the workplace involves the following steps:

- setting goals;
- analyzing the status quo—sometimes called the "baseline"—to determine how far the organization is from achieving its goals;
- planning for change (identifying actions needed to create change, and prioritizing them);
- putting necessary resources (including human resources) in place;
- implementing the actions identified in the planning process;
- addressing any obstacles or unexpected problems encountered in the change process;
- evaluating and reinforcing/rewarding progress; and
- setting new goals.

Each of these steps will be addressed in this chapter.

SETTING GOALS

Introduction

While much of the work of establishing or upgrading the workplace health and safety system will be carried out by middle management and the workers themselves, it can be a mistake to make these individuals responsible for setting occupational health and safety goals.

The reason for this is not that managers and workers are not equal to the task—quite the opposite. Workers observe process flaws, work risks, and near misses every day, and so they are uniquely well placed to identify safety problems. However, identifying problems that need solving is not the same thing as setting goals. A safety program that is based on managing problems tends to be reactive and focused on minimum compliance.

A reactive approach will achieve basic compliance, but in order to achieve superior safety performance, a proactive approach—one that involves setting ambitious goals and working toward them—is essential.

Setting organizational goals, identifying core values, and influencing the business culture is the responsibility of senior company management. To be effective, health and safety change must flow from the very top.

The Role of Senior Management

Responsibility for identifying and communicating the culture and values of an organization lies with the CEO or president and with the top ranks of management. Senior executives are expected to be accountable for a company's current performance, and to steer its future course. Successful businesses regularly assess whether their measurable results are predictive of success in meeting ultimate goals, and whether the company's performance is in line with its core values.

The kinds of results that a business can measure include sales/profitability, market share, rate of retention/loss of key employees, freedom from legal liabilities and regulatory sanctions/interference, and environmental performance. Safety performance also has a place on this list, because a company can compare its injury and illness rates from year to year as well as against industry competitors (for example, via incentive programs like NEER).

In areas where measurable results are disappointing, executives are expected to make adjustments designed to improve performance, and to be accountable to shareholders for the success or failure of these adjustments.

In the previous chapter, you learned that embracing safety as a core business value is one of the factors that sets safety leaders apart from other employers. For an individual, a core value is something essential with which the person self-identifies: it makes the person who he or she *is*. The same goes for a corporation—a core value is an intrinsic principle or standard that defines what the company is all about. It is even more important than a priority, because the concept of a priority implies that the priority is something that can be ranked—in some cases, *after* other priorities—and something that can change in importance over time. A common safety slogan is "safety is a priority." This statement affirms a company's view that safety is important. However, it falls short of the meaning of "safety is a core value." Making safety a priority makes it important; making it a core value makes it *essential*.

Only the CEO/president and the top executives have the authority to define a company's core values. For this reason, these individuals must assume leadership for the safety agenda, and must remain permanently engaged with that agenda. The authors of *Business Results Through Health and Safety* explain it the following way:

> Building and sustaining a culture and systems for safe operation is not a one-time effort. It is not a program. It is a way of doing business. It requires sustained effort. Management attention to health & safety in itself can produce dramatic improvements in accident, injury and illness rates. However, once management's attention is diverted to another issue health & safety results will quickly deteriorate. The way to sustain results is the implementation of a solid system for managing health & safety with on-going accountability for following the system and commitment to continuous improvement. (CME and WSIB, 2001, p. 58)

Once safety has been embraced as a core value of an organization, senior management can start to build what has been described as a "culture" of safety. Chapter 12 explains what is meant by a safety culture and explains how organizations can take steps to develop such a culture.

Building a System

As you learned above, a reactive response to health and safety at work tends to produce a minimum level of compliance. But there are other pitfalls of being reactive, instead of proactive, when it comes to health and safety.

In a reactive workplace, there is little if any effort given to the task of prioritizing OHS activities. If supervisors wait until workers point out risks before taking action to improve equipment or processes, the risks that will be addressed first will be those that affect the most assertive workers. Similarly, if managers institute processes—for example, a monthly OHS inspection of the manufacturing floor—in response to the passage of new legislative or regulatory provisions requiring those processes, the company will likely be a "late adopter" of the safety processes in question. This is true because compulsory (legislated) safety practices do not come out of nowhere. Instead, they very often evolve from the voluntary standards adopted by companies who are safety leaders.

Reacting to, rather than managing, safety challenges eliminates an employer's ability to determine its own priorities and to implement changes while making the most important use of limited resources.

So where do safety priorities come from? In well-organized workplaces, they are expressed in, and are generated by, a carefully planned health and safety system.

CSA-Z1000: A Blueprint for a Health and Safety System

CSA-Z1000 is a key workplace standard developed by the Canadian Standards Association (CSA). It provides guidelines for the creation of an occupational health and safety management system (OHSMS).

CSA-Z1000 is a voluntary standard, which means that choosing to follow it in developing your own OHSMS is optional. There are other models available, including many developed in the United States. Some companies may choose to hire a consultant to develop an OHSMS carefully customized to their business. Whatever approach you choose, and whichever model(s) you consult, the key issue to keep in mind is that the point of an OHSMS is to *simplify* the management of occupational health and safety in your organization. A system simplifies a management task by expressing a clear vision (for example, of the organization's values and goals) and by creating processes that make managing safety a part of regular business routines—not something that is dealt with in random fits and spurts in reaction to problems or legislative changes.

CSA-Z1000 describes its own purpose as follows (p. 1):

> This Standard specifies requirements for an occupational health and safety management system (OHSMS). The purpose of this Standard is to enable an organization to improve its occupational health and safety (OHS) performance, and thus reduce or prevent occupational injuries, illnesses, and fatalities, by:
>
>   (a) establishing, maintaining, and improving an OHSMS that will identify and then eliminate or control all OHS hazards and risks;
>   (b) ensuring conformity with its OHS policy; and
>   (c) demonstrating conformity with this Standard by
>
>   (i) making a self-determination and self-declaration;
>   (ii) receiving confirmation of its self-declaration by an outside party, independent of the organization; or
>   (iii) gaining certification/registration of its OHSMS by an outside organization.
>
> All the requirements in this Standard are intended to be incorporated into an OHSMS. The extent of the application will depend on the circumstances particular to the organization, such as the nature and location of its operations, and the conditions in which it functions.

The standard incorporates the following elements:

- a statement of its purpose;

- a description of the scope of the application of the standard (in this case, the standard claims to be applicable to "an organization of any size or type");

- a list of reference publications (in this case, the list contains the standard numbers of six other CSA standards that relate to particular aspects of occupational safety, for example, CAN/CSA-Z731-03, Emergency Preparedness and Response);

- a list of definitions for terms used in the text of the standard (for example, "incident" is defined as "an occurrence, arising in the course of work, that could result in an injury or illness (includes near misses)").

The standard then goes on to provide a series of numbered guidelines for the development of an OHSMS, and three Annexes (similar to appendices) that help inform these guidelines.

The guidelines established by standard CSA-Z1000 fall under two general headings: "Occupational Health and Safety Management System" and "Management Review and Continual Improvement."

Under the first heading the topics addressed include:

- a suggested process path for addressing OHS with key steps summarized under the headings "Plan," "Do," "Check," "Act";

- a statement of the importance of appropriate commitment to, and leadership for, OHS activities. For example, subsection 4.2.2.1 of the standard provides as follows:

    Senior management shall provide leadership for OHS activities and assume overall responsibility for the OHSMS. This responsibility includes:

    (a) establishing, actively promoting, and maintaining the OHSMS;
    (b) providing appropriate financial, human, and organizational resources to plan, implement, check, review, and correct the OHSMS;
    (c) defining roles, assigning responsibilities, establishing accountability, and delegating authority to implement an effective OHSMS;
    (d) establishing and implementing an OHS policy and measurable objectives;
    (e) reviewing the organization's OHSMS at planned intervals;
    (f) ensuring that workers and worker representatives are consulted as required by this Standard; and
    (g) encouraging active participation on the part of workers and worker representatives in the establishment and maintenance of the OHSMS.

- a statement about the importance of planning to the success of an OHSMS, and suggested topics about which planning may be needed (for example, reviews, legal compliance, hazard identification, and OHS objectives and targets);

- guidelines for the implementation of an OHSMS, with specific recommendations with respect to preventive and protective measures;

emergency prevention, preparedness, and response; competence and training; communication and awareness; procurement and contracting; change management; and documentation, control of documents, and records management. As an example, one recommendation in the "Management of Change" portion (subsection 4.4.7.1) provides as follows:

> The organization shall establish and maintain procedures to identify, assess, and eliminate or control OHS hazards and risks associated with:
>
> (a) new processes or operations at the design stage;
> (b) significant changes to its work procedures, equipment, organizational structure, staffing, products, services, or suppliers;
> (c) developments in OHS knowledge and technology; and
> (d) changes to OHS legal requirements and, where applicable, collective agreements or other requirements. Such procedures should include information sessions and training for workers as appropriate.

- guidelines for evaluation and corrective action, including in the areas of monitoring and measurement of OHS performance, incident investigation and analysis, internal audits, and prevention and corrective action.

Under the major heading "Management Review and Continual Improvement," the standard addresses the following topics:

- senior management's ongoing duty to review the OHSMS at regular intervals to ensure "its continuing suitability, adequacy, and effectiveness";

- the need for a commitment to continual improvement when it comes to OHS;

- guidelines for the content of management "review input" and "review output," for example, under heading 5.4:

> Action plans shall be developed from the management review. The organization shall have a process for recording and communicating the findings, conclusions, and action plans of the management review to the persons responsible for action and the workers and worker representatives.

CSA-Z1000 is an excellent starting point for new employers, or for those who are motivated to revitalize an existing occupational health and safety program. A copy can be ordered from the CSA (www.csa.ca).

ASSESSING THE STATUS QUO

Introduction

Before improvements to occupational health and safety can be prioritized and planned, a business needs to know how the realities of its safety program compare to the goals it has set for itself. Assessing the safety status quo is a complicated but necessary task, and should be handled systematically.

There are many resources available to support employers in assessing the risks, safeguards, and policies in place in their workplace. Appendix G of this book contains a sample list of resources for JHSCs, and these resources are equally useful to managers looking for guidance in their assessment efforts. The WSIB document *Business Results for Health and Safety* provides a particularly detailed, comprehensive framework that employers can use to rank their workplace's performance in many areas of safety.

The following sections highlight steps that should be taken toward developing an understanding of the organization's safety status quo (baseline).

Reviewing Injury and Accident Statistics

Every employer is required, under the OHSA and the WSIA, to report certain incidents to the WSIB and/or the MOL and to keep records of others. Well-organized employers will choose to keep internal records of *all* accidents and near misses (incidents in which a person narrowly escapes injury), whether or not an employee has suffered a lost time injury.

Where an employer is enrolled in an incentive program like NEER, CAD-7, or MAP, its injury and illness statistics will be used to determine premium rebates or penalties. These programs adjust premiums based on the accident rates of all the businesses in an employer's class, and information about average rates is readily available for comparison purposes.

Reviewing both the company's internal incident statistics and the general industry statistics available from the WSIB provides a useful starting point for determining the baseline rate of accidents, and for determining whether the company has improved or faltered when compared to past years.

PHYSICAL INSPECTION OF WORKPLACE

A physical inspection of the workplace is completed in an effort to identify safety hazards, aspects of the workflow that are awkward, deficiencies in equipment, vulnerability to external disruptors (for example, weather), and unhealthy conditions such as overcrowding, unacceptable working temperatures, poor ventilation, low light, and exposure risks.

Hazard identification is an expert skill. The physical inspection of the workplace should be conducted by an employee or employees who have been trained in hazard identification, or by a safety specialist consultant hired for the purpose. If it has not already been done, the person conducting the hazard identification inspection should develop and work from a checklist designed to ensure thoroughness and continuity. Sample inspection checklists for various kinds of workplaces are available from many different sources, including:

- the MOL;
- the CSA;
- SafeCanada.ca (a federal government site that provides links to various government resources);
- the IAPA;
- the CCOHS; and
- industry-specific safety associations, such as OSACH.

The authors of *Business Results Through Health and Safety* suggest that it is a good idea to have a team of at least three people conduct a workplace inspection together. This approach tends to even out individual bias, for example, in situations where one inspector rates a risk as low and another rates it as high.

In deciding what needs to be inspected, it can be useful to refer to detailed regulations, such as the Industrial Establishments regulation or the Health Care and Residential Facilities regulation made under the OHSA. Inspectors should also take into account relevant industry safety standards to determine whether the employer's equipment, layout, and processes meet these standards.

Inspectors should not hesitate to speak to workers when conducting their inspections. Often, workers identify risks long before their supervisors do, and can report on problems that happen in their work that may not be obvious at the time of an inspection. For example, an inspection of an outdoor worksite may be conducted on a windless day, and may not reveal unacceptable vibration of machinery in moderate wind conditions.

Reviewing Existing Policies and Procedures

Once the physical workplace has been assessed, the procedures and policies in place in that workplace should be reviewed. In a small workplace, there may only be, if anything, a general OHS policy. In a large industrial business, there may be written procedures for a wide range of tasks, functions, and activities, including:

- the use of particular kinds of equipment;
- the procedure for machinery lockouts;
- the use of personal protection equipment (PPE);
- the handling of specific risks (for example, high temperatures);
- the administration of first aid in the event of accidents;
- responding to a work stoppage or refusal; and
- reporting incidents/accidents.

In reviewing policies and procedures, it is important to consider not only the written content of the procedure, but also compliance. Is the procedure well understood? Is it being followed? Is it considered to be effective, cumbersome, or inadequate? A procedure is only useful if it is being used as intended by its creators.

The agencies mentioned above as being sources of inspection checklists may also have sample or model policies and procedures to govern OHS functions. In addition, the WSIB offers two "audit" documents that are designed as tools to help a business assess the quality of its policies. See the *Compliance Audit for Employers* document (available from the WSIB) and the *Core Health and Safety Audit* document (available from Workwell, a WSIB program).

The process of creating policies and procedures should never be considered "complete," but rather as ongoing, with opportunities for continuous improvement. Each change in equipment or work processes should trigger the development of new safety policies to manage new risks. For this reason, a policy analysis should examine not only the content of existing policies, but also the business's track record when it comes to the policy development process. Is there a long lag between process or equipment changes and the introduction of new policies? Is there a schedule in place for policy review, or is policy development reactive, with most policies being developed in response to safety incidents?

In general, the creation of a new overall OHS policy will be one of the last steps in an employer's effort to revamp its health and safety program. The new policy will be designed to reinforce the progress made, the commitments undertaken, and the company's expectations for its future performance.

Evaluating Culture, Perceptions, and Behaviours

This book has already mentioned the impact of organizational culture on health and safety. Having a "culture of safety" in place at the worksite makes all other OHS initiatives easier to implement.

Often, the processes of hazard assessment and a policy analysis—the two steps already mentioned—will provide some valuable insights into the business's safety culture (or lack thereof). To further pinpoint areas of concern, it is useful to consult with supervisors in each department or area. Senior management may ask supervisors to identify any compliance obstacles with which they have had recent or ongoing problems.

Worker behaviours are an aspect of the "human dimension" of OHS. Employees can be supportive in theory of the goal of safe work, but their behaviours may still be unsafe. This inconsistency between beliefs and behaviours can have many causes, for example:

- equipment deficiencies (having to work with dangerous/outdated/ unguarded equipment; lack of PPE);

- insufficient training (workers may not know *how* to work safely);

- overwork (due, for example, to understaffing);

- unrealistic productivity targets (which can force employees to rush and cut corners);

- incompetent supervision (for example, having supervisors who fail to enforce safety procedures); and

- underestimating risks (for example, if a worker has skipped a safety step 99 times, she may perceive little risk in skipping it the 100th time—and yet the risk may be very real).

Again, the best sources of information about behaviours are workers and supervisors. These individuals should be asked to speak openly about what really happens at the worksite.

Another source of information about problem behaviours are the company's accident reports. An analysis of these reports can easily pinpoint areas of particular concern. These records will be even more valuable if the employer encourages workers and supervisors to report near misses.

Evaluating Formal Compliance and Reporting

Finally, the employer should assess how it has been performing with respect to formal compliance and reporting responsibilities. These responsibilities include reporting accidents to the WSIB and/or the MOL on a timely basis, reporting work stoppages, responding to MOL requests for information, complying with MOL orders, and completing filings required under programs such as NEER.

Another key aspect of formal compliance is the state of the employer's performance in its internal relations, either with employees or with its own Health and Safety Representative or JHSC. Has the employer been investigating work refusals and stoppages in a timely manner? Has it been responding, within the legislated timeframe, to JHSC reports (for example, the OHSA requires that the employer respond within 21 days to a JHSC's investigative report on a serious accident)?

Failure to meet deadlines for administrative compliance responsibilities can be a sign that the employer's management of safety is inefficient or chaotic, or that the employer's administrative department(s) are understaffed.

PLANNING FOR CHANGE

Once a business has successfully engaged its senior executives in setting safety goals and adopting safety as a core value, and once a thorough assessment of the safety status quo has been performed, the next step is to plan for change.

Adopting a proactive approach to safety promotion means that the business has decided to be governed by its own OHS priorities, instead of simply being reactive. The baseline analysis of the company's OHS "fitness" will have revealed areas of concern, and it will now be up to management to prioritize those concerns. What should be tackled first?

There are many ways to prioritize tasks, and each employer will have to determine its own. In general, however, before addressing more theoretical issues (for example, an overhaul of written procedures), the business will need to remedy any serious hazards identified by the physical inspection. If the resources available for attending to safety are limited, money should first be devoted to replacing dangerous equipment and fixing physical plant hazards (for example, problems with the electrical system).

After immediate hazards have been eliminated, a good second priority will likely be the elimination of any gaps in employee safety training. Many accidents happen (especially accidents involving young workers) because of a lack of basic safety training.

Eliminating immediate physical hazards and dangerous training gaps can be considered the "first aid" stage of a process of safety program renewal. Once first aid has been administered, the company can move on to address other areas of concern in a priority order of its own choosing. Tasks remaining after safety review "first aid" has been administered might include, in no particular priority order:

- addressing regulatory compliance gaps; for example, establishing (and training) a JHSC, and/or bringing communications and procedures into compliance with the WHMIS;

- addressing internal compliance gaps and administrative inefficiencies (for example, by allocating administrative (usually, human) resources to clearing out any backlog in internal record keeping or management, outstanding JHSC recommendations requiring responses, or worker WSIA cases needing management/follow-up);

- creating a training program and schedule that ensures that all workers and supervisors are appropriately trained at all times, that training is refreshed, and that training is provided in a timely way to address specific situations (for example, the introduction of a new hazardous substance, the adoption of a new process, or the return to work of a worker who was off on compensation);

- developing and implementing a protocol to govern and document the investigation of workplace incidents;

- introducing supportive programs such as a WHP program (these were introduced in chapter 7) and/or an EAP;

- introducing an enhanced reporting system that will allow the employer to analyze near-miss incidents and other incidents without injuries, to measure the benefits of safety initiatives, and to measure the full complement of costs of injuries; and

- creating incentive programs designed to support a culture of safety, at either or both of the worker and management levels.

These are examples only; the specific tasks undertaken by an individual employer should be dictated by the priorities that emerge from the processes of management recommitment to OHS and the baseline analysis of the safety state of the workplace.

An effective plan for change will allocate responsibility for each aspect of implementation to competent individuals, and will incorporate deadlines for the completion of the new initiatives.

MARSHALLING RESOURCES

Before the plan developed in the previous step can be implemented, the organization will need to have appropriate resources in place. These resources can be financial—for example, funds to replace outdated personal protective devices—physical—for example, training manuals, posters, MSDSs for WHMIS compliance, and prizes/plaques/rewards for use in an incentive program—and, finally, human resources.

Building a budget for a safety program should be a task assigned to an executive who has the authority to spend company money. Determining what is needed, in terms of resources, to support the new initiatives may be a task that needs to be split up across multiple departments.

The most crucial resource of any safety program, however, is people.

Both management and workers should already perceive themselves as having general OHS responsibilities: the OSHA creates an IRS that assigns OHS roles to the employer, supervisors, workers, union representatives, and other workplace parties.

In a workplace that is even minimally compliant with the OHSA, there will be individuals who already have specific OHS responsibilities: there will be either a Health and Safety Representative (in small workplaces) or a JHSC. It is important to recognize that these individuals already have significant responsibilities under the OHSA, and that it is not realistic to expect them to undertake a slate of additional responsibilities in support of the employer's safety program development or renewal initiative.

Larger workplaces, or those with significant safety hazards, may have other individuals with safety responsibilities. These may include:

- supervisors and managers with specific safety-related inspection or oversight functions (for example, forepersons who are expected to inspect the worksite and/or equipment at the beginning of a shift, or who are expected to supervise machinery shutdowns or repairs);

- supervisors or managers who are charged with conducting investigations after safety incidents;

- supervisors or managers charged with human resources aspects of workplace health and safety (for example, processing WSIA claims and managing follow-up);

- supervisors or managers charged with OHS administration and/or compliance functions (for example, managers who serve as the employer's liaison with the MOL or other regulatory agencies);

- trainers who deliver OHS, WHMIS, and other safety training;

- workers or managers with first aid training who are designated as first aid providers in the wake of an accident;

- nurses, counsellors, or occupational therapists charged with providing health-related services;

- workers and managers who work in a department specifically charged with occupational health and safety management (this department may employ specialists, for example, occupational hygienists).

One aspect of the process of safety program development or renewal should be a review of the roles listed above. This will allow the employer to determine whether it has human resources appropriate to its needs, and whether there are gaps that should be filled. This exercise will require a significant amount of judgment, and considerations will include the following:

- Are aspects of OHS support (training, for example, or psychological counselling) best delivered in house, or by referral of employees to outside providers? (Providers can include, for example, agencies that specialize in OHS training, or an EAP that provides health care services);

- Is an aspect of OHS support best met by an employee who will play the role on a permanent basis, or by a consultant engaged to address a specific, temporary need?

Examples of needs that may be best addressed by a consultant include:

- a thorough baseline inspection of the workplace to identify OHS problems and priorities;

- an inspection and/or planning process in support of a particular compliance obligation (for example, the preparation of a Pre-Start Health and Safety Review or participation in a Workwell audit; these were discussed in chapters 6 and 7, respectively);

- the investigation and remediation of a particular OHS problem (for example, poor air quality in a building); or

- the delivery of a new and comprehensive training program (which may include "train the trainer" aspects designed to permit the employer to deliver refresher programming in house).

In assessing the sufficiency of human resources to support occupational health and safety, it is not enough for the employer to consider whether there are *enough* employees committed to OHS. The employer must also consider the technical expertise of its workforce as compared with the technical demands of its OHS review. In some cases, complicated OHS activities will require either that existing employees obtain special training, or that outside experts be retained, at least on a temporary basis.

☼ Success Story

Retaining People by Investing in Safety Technologies

In the 1990s the demand for personal care workers was high across North America. One Seattle, Washington nursing home described the labour market for these workers as "at crisis point." Nursing homes, however, tend to be dangerous places to work. In 1998, the United States Occupational Health and Safety Administration ranked nursing homes in the top ten sources of non-fatal workplace accidents, with 13.8 injuries or illnesses per 100 full-time workers. (United States Department of Labor, 1999, p. 12). (Canadian healthcare facilities are also plagued with higher-than-average injury rates.)

One of the biggest contributors to the high rate of injuries among nursing home staff is the need to lift and move patients. Back injuries account for 42 percent of all injuries to workers in these settings.

Armed with these statistics, the Bessie Burton Sullivan Skilled Nursing Residence in Seattle knew that if it was going to retain healthy staff in the midst of a very tight labour market, it was going to have to prevent lifting-related injuries. The facility invested $22,000 in new ergonomic lifting equipment. Because a back injury in Washington cost an average of $6,103 in medical and wage-replacement expenses at the time, the nursing home knew that if more than four average injuries were prevented, they would recoup their equipment costs in a single year. In fact, the results of the facility's investment in injury-reducing technologies and practices was much *more* cost effective than that: the nursing home, which had 73 lost-time days from lifting injuries in 1997, had only 7 in 2000.

The Bessie Burton Sullivan experience is a clear example of how a facility can protect its human resources and save money by investing in safety technology.

IMPLEMENTING CHANGES

Introduction

Once upper management has set OHS goals for the organization, a baseline OHS assessment has been completed, immediate OHS problems have been addressed (on a "first aid"-style basis), and the required resources (including personnel resources) have been put in place, the organization is ready to begin the process of implementing the changes dictated by its identified priorities. The following sections introduce some of the more common OHS activities that may need to be implemented or managed; not all employers will need to address of all these. They are introduced simply because they are the most common areas requiring attention.

Health and Safety Representative or JHSC

The employer may need to review the application of the OHSA to its workplace and address basic compliance issues. For example, if there is none, the employer may need to appoint a Health and Safety Representative or create a JHSC, and arrange for training and certification of members as appropriate.

As explained in chapter 3, at least one member on each of the management and worker sides of a JHSC must be certified. This means that the members must complete training approved by the WSIB.

There are two levels of certification. Basic certification training focuses on OHS law, general hazard awareness and identification, investigative techniques, and prevention strategies. The training itself is delivered by providers approved by the WSIB. The WSIB maintains a list of these approved providers. Basic certification training is currently available from providers in about 30 communities in Ontario.

The second level of training is hazard-specific training. This training—which is also compulsory for the purpose of certification—is focused on particular risks in individual workplaces.

Identifying what is needed, in terms of hazard-specific training, is the responsibility of the employer. The employer is required to assess the hazards that exist in the workplace and to identify the training required to address those hazards. Once this process is completed, the employer must fill out a Workplace-Specific Hazard Training Confirmation form, provided by the WSIB, that outlines these training needs. This form is reproduced in appendix H.

In some cases, delivery of this training is organized in sector-specific ways, by combining coverage of multiple hazards typical in, for example, construction, health care, or police services. This training can be delivered by sector-specific health and safety associations, for example, the Construction Safety Association of Ontario (CSAO); by private consultants; or by the employer's own in-house trainers (who have presumably gained significant expertise as far as the hazards characteristic of the workplace are concerned).

Once the JHSC is appointed, trained, and certified, it will be ready to discharge its responsibilities under the OHSA.

WHMIS Compliance

Whenever hazardous materials are used in a workplace, the workplace must ensure that it complies with the WHMIS (discussed in chapter 4).

The employer should designate a manager or supervisor (or more than one) as responsible for managing WHMIS compliance. That manager may need to conduct a review of the WHMIS procedures and materials in place in the workplace to determine whether materials are labelled correctly and whether up-to-date MSDSs are available.

The person in charge of WHMIS compliance should also review the system's actual performance in the workplace. Do employees understand WHMIS labels? Do they know how to find MSDSs? Do they follow instructions for material handling and storage? If the review reveals problems, refresher training should be scheduled to reinforce proper procedures.

The final step in bringing the workplace into WHMIS compliance is to establish a schedule for regular reviews of the system and updating of materials.

Scheduling Inspections

The employer's ability to manage hazards depends on its capacity to identify these in the first place. A regular schedule of OHS inspections is essential for workplaces of any size.

Managers or supervisors should be given specific responsibility for workplace inspections, and these responsibilities should be given only to those employees who are appropriately trained in hazard investigation.

The frequency of inspections will vary according to the type of workplace and the hazards present. A business office may require a brief walk-through once a month, and one thorough annual inspection. A manufacturing floor, by contrast, should probably be inspected by forepeople at the beginning of every shift, and by a safety manager at least monthly.

Inspectors should develop and work from a checklist. Checklists ensure that nothing is overlooked and, if filed for future reference, allow inspectors to pinpoint the onset of problems. Many resources are available to help inspectors develop checklists that are appropriate to particular kinds of workplaces, and even particular kinds of equipment. Appendix G lists resources that may be of use to a JHSC; the same resources can be of assistance to anyone who is charged with inspecting a workplace.

Developing and Renewing Policies and Procedures

A baseline risk assessment and a process of ongoing inspections will bring predictable risks, in the workplace, to the attention of management. Often, the best way to manage these risks is by developing policies and procedures designed to assist employees in working safely.

Safety policies and procedures can be as simple as a requirement to clean up all spills immediately. Where the situation requires it, they can also be much more complex; for example, a large department store's evacuation plan is a safety procedure, as is a bus station's protocol for dealing with bomb threat phone messages or a hospital's guideline with respect to dealing with violent ER patients.

Some workplaces will have only a handful of procedures, while others will have dozens. Often, the employees best placed to conduct reviews of procedures (and to prepare updates, where needed) are supervisors or managers of specific departments or areas. Someone more senior than these individuals may also be charged with the duty to ensure that policy review happens on a regular basis across the organization as a whole.

Models and precedents for safety procedures abound; appendix G includes some possible sources.

One general kind of policy merits mention here. Studies show that accidents are more common shortly after there has been a change in tools, equipment, work environment, or personnel. The reason for this is that workers may be unaware of new risks introduced as a result of the change. To minimize accidents related to changes in the workplace, each introduction of a new piece of equipment, a new process, or a new worker should trigger an automatic safety protocol that includes both an inspection of the altered work area and training for the workers. A policy that makes safety planning a part of every new acquisition or hire will be an excellent investment in the company's safety profile.

Preparing an overall OHS policy may be an appropriate final step in the process of safety program development or renewal. The overall policy should incorporate, at minimum:

- the company's attitude toward safety (as noted elsewhere, employers are urged to make safety a core value of the organization);

- the company's safety-related goals (for example, to reduce the rate of injuries or illnesses, and perhaps by how much and by when);

- a commitment by senior executives to safety;

- a statement about the responsibility of other workplace parties for safety;

- a summary of known hazards and a plan for minimizing risks;

- a training plan;

- a plan for evaluating safety initiatives and programs; and

- a plan for ongoing review and continuous improvement.

A good resource to use as a starting point in developing an occupational health and safety policy is included in Appendix A to the MOL's *A Guide to the Occupational Health and Safety Act* (available on the Ministry's website).

FACTS AND FIGURES

Bracing for Pandemic Flu

Instances of bird-to-human transmission of influenza type A, H5N1 virus ("avian flu") in the period since 1997 have led epidemiologists to point out that North America is "overdue" for a flu pandemic. A pandemic is an international epidemic of disease that affects a large proportion of the population. Pandemics typically arise about once every 10 to 15 years, when a new strain of a virus for which people do not have immunity mutates and is transmitted to humans. The most serious flu pandemic this century happened in 1918, while the most recent occurred in 1968. Most experts expect avian flu to be the culprit in the next pandemic.

Pandemic flu differs from "normal" flu in that the population's lack of immunity allows the disease to be transmitted from person to person very easily. Symptoms tend to be more severe than with the common flu. While the elderly, the very young, and the immunocompromised tend to be at greatest risk from "normal" flu, pandemic flu produces severe symptoms even in healthy adults, and results in a rate of mortality much higher than that seen in "normal" flu outbreaks. (For example, more than half of the people diagnosed with influenza type A H5N1 have died. However, it is important to remember that milder cases of the virus have likely not been reported or documented.)

A flu pandemic can last up to two years, with illness spreading in multiple waves that last a few weeks, separated by periods of up to three to nine months. Experts predict that if a flu pandemic hits, workplaces may have 15 to 50 percent of their workforce off sick or caring for sick relatives at the height of the epidemic, and that 1 percent of infected individuals will die.

Some workplaces will be forced to close, and key resources such as food, hydroelectric power, and water may be interrupted.

Companies that would suffer serious losses if they were forced to close should act now to create a business continuity plan. Examples of these kinds of plans are available from many sources, including Canadian Manufacturers and Exporters (CME), which offers a resource called *Influenza Pandemic: Continuity Planning Guide for Canadian Businesses*.

Other pandemic-planning initiatives employers may wish to take (besides creating a continuity plan) include:

- revising sick leave and family leave policies with pandemic influenza in mind;
- providing reading materials for workers, for example, posters about hand hygiene in washrooms (posters and other educational materials that deal specifically with influenza can be obtained from sources like the CCOHS); and
- developing a web page or portal devoted to the subject of pandemic flu; the page can be restricted to worker users and can contain well-chosen resources and links, information about employer policies, and general information about prevention, treatment, and how to care for ill family members.

The employer may also wish to obtain training for key employees about pandemic-related issues, for example, the benefits of PPE, how to address worker fears, effective referrals, and topics for employee training (for example, how to use social distancing as a way to reduce transmission once an outbreak has begun).

Managing Training

Adequate training for workers is perhaps the single most important feature of a good workplace safety program. Both general training (training in how to do one's work tasks) and safety-specific training play a role in reducing workplace injuries and illnesses. As noted in the previous chapter, the especially high rate of injuries to younger workers (25 and under) has been shown to be associated, in many cases, with deficiencies in training. Training makes an important difference, which is the reason why it is a central requirement of programs, such as the WHMIS, that are aimed at reducing identifiable risks.

Safety training should be delivered to the following individuals on the following occasions:

- to new hires as soon as possible after their start date (or in industrial settings, before any work with machinery is done);
- to workers returning from an extended leave of any kind;

- to existing workers who are taking on new duties or who have been promoted to supervisory positions;
- to all workers who handle hazardous materials as soon as they are required to handle them, or immediately after introducing a new material;
- whenever there is a change in work setting, equipment, or processes; and
- whenever there is something new to learn (for example, when the employer learns of a new industry best practice that can help reduce accidents in the workplace).

"Refresher" training should also be offered to all workers at regular intervals. The content of this training should be kept up to date by a training manager who regularly reviews innovations in the field of occupational health and safety.

Some employees will require specialized training. As explained elsewhere in this text, members of a JHSC need training in how to identify workplace hazards as well as in other areas. Safety supervisors will need training in a wider range of areas than will front-line workers, and safety trainers will need training in how to design and deliver training.

Employers should pay particular attention to the training needs of young workers because of their increased vulnerability to injury. There are many tools available to assist employers in protecting young workers, including the Passport to Safety education program (see www.passporttosafety.com).

Besides offering a thorough and well-organized training program, employers should take steps to create an organizational culture in which employees are encouraged to ask questions about safety, challenge worrisome practices, and request training in areas that they themselves have identified. An example of a way in which this culture might be reinforced is by the employer offering incentives—perhaps a quarterly award of some sort—for a worker "safety innovator."

When developing training, very special attention should be paid to the training given to first-line supervisors. These individuals play a key role when it comes to communicating organizational culture to workers. In *Leading with Safety*, Thomas R. Krause notes that:

> In most organizations the first-line supervisor is a key influence on organizational effectiveness. The supervisor links management and the workforce. Supervisors ... often have more credibility with workers than managers. Workers look to the supervisor to interpret organizational priorities and changes and they may ask the supervisor directly about the meaning of some particular management action. More typically, the supervisor interprets management actions in light of his or her experience and passes them on in the form of directions, informal comments, and reactions to workers' actions. In many respects workers take the words and deeds of the supervisor to represent "the company." That is, worker perceptions about the organization are filtered through the supervisor. (Krause, 2005, p. 30)

Because of supervisors' key role in communicating and shaping organizational culture, supervisor training should incorporate well-designed messages about the organization's commitment to the core value of safety. These messages should be reinforced with appropriate management attitudes in the context of supervisors' interactions with managers and senior executives. A supervisor who is trained to encourage workers to report problems but who is met with resistance when she reports problems to her own superiors will convey mixed messages about safety to the workers under her authority.

> ## Thinking Safe
> ### Play 'n' Learn with Prevent-it.ca
> For a fun and informative introduction to identifying workplace hazards, visit www.prevent-it.ca. This site, sponsored by the WSIB, offers, among other learning tools, an interactive hazard-spotting challenge. Explore the businesses on a commercial/industrial street and see if you can identify the workplace hazards.

Introducing Health Promotion Programs

Once a company has accidents and basic OHS compliance under control, it can turn its attention to improving safety performance. Above-average safety performance will result in premium savings under incentive programs, and can have an impact on overall productivity. How does a company go from competence to excellence when it comes to health and safety? Prevention and promotion programs can help.

Health promotion programs were mentioned in chapter 7. These programs, which often focus on common health issues like smoking and sedentary lifestyles, are based on the premise that the workplace is a convenient environment in which to educate people. With fairly limited effort and expense on the employer's part, employees can be given access to health information that they might not otherwise receive. This information can come in the form of educational booklets and self-assessment tools, or in the form of lunchtime seminars or classes that employees can attend on a voluntary basis. For example, if the number of interested employees warrants it, the Weight Watchers weight management organization can be engaged to run a weekly lunchtime session at a workplace. An employer can also offer one-time seminars on general lifestyle topics, such as "Coping with the Needs of Aging Parents" or "Cutting Your Heart Disease Risk."

These programs have been demonstrated to be effective in terms of real dollar savings in many contexts. Cost–benefit analyses aside, however, the employer's willingness to research, arrange, and deliver this kind of programming can send a powerful message about the sincerity of the company's commitment to employee well-being.

A more comprehensive health-related program with which most human resources managers are familiar is the EAP. An EAP is often a feature of the benefits program offered to employees. It may provide access, for employees and sometimes members of their immediate families, to health and wellness-related services such as psychological counselling and financial services. Some EAPs provide access to health services like nutritional counselling or massage therapy, but it is generally not cost-effective to the employer to offer free access to services that the employee would otherwise be able to access under his or her provincial health insurance. Typical areas in which an EAP can provide help include:

- job stress;
- relationship problems;

- eldercare, childcare, parenting issues;

- harassment;

- substance abuse;

- separation and loss;

- balancing work and family;

- financial or legal problems; and

- family violence.

EAPs are often delivered by a third-party provider—a company that specializes in staffing, designing, managing, and delivering these programs for a range of clients. Contracting for EAP services in this way eliminates the need for the employer to assume the responsibility for recruiting and screening an appropriate complement of service providers.

For help in choosing an appropriate EAP provider, employers may wish to consult the CCOHS, which offers a free resource containing tips about how to choose an EAP on its website. Employers will also want to review the website of the Canadian Employee Assistance Program Association at www.ceapa.ca.

A third kind of health and safety "program" in place in many organizations involves offering health services in the workplace, for example, by having a nurse or physician on staff. While all employers should be able to deliver first aid—via properly-trained and designated first-aiders and a well-stocked first aid kit—the rate of injuries and illnesses in some settings warrants a full-time employee (or more than one) dedicated to managing the immediate health needs of workers. The most common choice of provider is a nurse, though some larger employers employ physicians, mental health professionals (counsellors, therapists), and/or occupational therapists. Hospitals tend to be leaders in this area, probably because they are already in the business of health care delivery, but also because their employees face particularly high health risks.

Where the workforce's typical first aid needs are modest, a program that might make sense will be one in which there are one or more designated first-aiders (ideally with CPR training) available on each shift, a small treatment room or area, and an appropriate supply of first aid supplies. The first-aiders may also be encouraged to inform themselves about health care services in the community, so that they can refer workers to providers on request. First-aiders will need thorough training on the subject of protection of workers' personal privacy and private health information. They should also receive training with respect to their role in supporting accident investigations.

Sophisticated health services programs can include health surveillance programs that are designed to screen employees for health problems associated with the most prominent hazards in the workplace. For example, most hospitals conduct ongoing general health surveillance programs and, in the event of an outbreak (for example, of influenza) a specific, temporary surveillance program may be instituted to ensure that employees with symptoms are screened and that those diagnosed are excused from contact with patients and receive appropriate treatment.

Managing Investigations

Despite an employer's best efforts, it is likely that the occasional accident will happen. The employer's response to workplace accidents will often determine whether or not the accident is likely to be repeated in the future, and what impact it will have on worker morale.

Post-incident compliance is the subject of the next chapter. However, in order to be able to respond effectively to accidents, the employer should be prepared with a procedure to guide investigations. Specific individuals should be charged with the duty to investigate accidents, and they should have a clear understanding of what they should be collecting in terms of information, what they should be recording for future reference purposes, and what they will need to report to various authorities (for example, the MOL, the Ministry of the Environment, any other government agency that regulates the employer's industry, and/or the WSIB).

There are many resources available to assist in developing an investigation procedure. One example is the step-by-step plan included in *Developing an Effective Safety Culture: A Leadership Approach*, by James E. Roughton and James J. Mercurio.

Investigations will be discussed in greater detail in chapter 9.

MOTIVATING, REWARDING, AND EVALUATING PROGRESS

Introduction

Once the process of safety program development or renewal has been completed, the employer should not allow the subject of worker health and safety to drop off the company's radar. In order to achieve excellent safety performance, a business must adopt a commitment to continuous improvement of its safety initiatives. Continuous improvement can only be achieved when an employer can keep the entire staff—from senior executives right down to front-line workers—motivated.

Motivation

Many employees today are under an incredible amount of pressure at work. Particularly where safety-related functions are not the central focus of a worker's job, it can be a challenge to persuade the worker to devote energy and time toward responsibilities that are not directly tied to his or her business identity or function within the company.

One of the simplest ways of reinforcing the importance of safety responsibilities is to ensure that workers view these as part of their job description. This view of safety can be emphasized at the hiring and training stages, but many employers forget to reinforce safety responsibilities when it comes time to evaluate performance. By defining explicit safety goals for every job description and measuring performance by reference to these goals at the same time as other facets of the work are measured, the employer can send the message that safety really is *everyone's* job.

Another way to motivate workers to achieve safety goals is to encourage them to take ownership, at least in part, of the safety program. Throughout the steps discussed in this chapter are opportunities to seek worker input on safety problems and solutions. By encouraging workers to protect themselves and their peers by helping to shape the safety agenda, the employer can motivate voluntary—and even enthusiastic—worker participation in safety initiatives.

Finally, workers can be motivated to support safety initiatives via incentive programs that reward performance. Incentive programs can encourage both individual and team-based participation; they will be discussed further in chapter 12.

> ## Thinking Safe
> ### Choosing a Health and Safety Month
> Routinizing the process of health and safety renewal can help prevent safety from falling "off the radar" for management. The authors of the *HR Manager's Guide to Health and Safety* suggest choosing a "health and safety month" for the workplace:
>
> > We recommend that employers pick one month per year to be the safety month. Employers can then use that month to review and revise their health and safety policy and provide updated training to workers. (Knight et al., 2004, p. 19)

Evaluation

Evaluation of the workplace's performance in implementing a new or improved safety performance will require, for comparison purposes, an assessment of the same areas reviewed in the course of the initial baseline review (suggested earlier in this chapter). The timing of this evaluation will depend on the size of the workplace and the degree of change mandated by the planning process. Where significant changes were required, the review timing should allow for enough time for implementation; however, delaying a review for too long can allow the initial motivation, on the part of the stewards of the renewal, to dampen. In most cases, a post-implementation review should not be delayed longer than six months from the beginning of the renewal/implementation process.

Where the review identifies areas in which improvements were delayed, it is important to investigate the reason for the delays. Were unanticipated obstacles encountered? Were the program goals poorly understood? Were improvements stalled by insufficient resources? These obstacles should be viewed as part of the learning process. Once obstacles to safety progress are understood, management will be able to take steps to address them.

> ## HARD LESSONS
> ### Where's Your Offhand?
> Solving persistent safety problems often requires more than just shopping for equipment guards or PPE. Robert Pater, a Portland, Oregon-based safety consultant, notes that it is important to consider not only equipment features, but also worker behaviour. Pater reminds us that worker behaviour includes not only conscious risk-taking behaviour on the part of workers, but also unconscious behaviours.

> In an article published in 2005, Pater referred to the common workplace problem of hand injuries. He noted that most manual workers use their non-dominant "offhand" hand much less often than their dominant hand. A worker who is focused on perfoming a task with his dominant hand may often be unaware of what he is doing with, or where he is placing, his offhand. This can mean that the worker may allow his offhand to fall into a pinch point, touch a hot surface, or be injured in one of many possible ways.
>
> Simply reminding manual workers to pay attention to what they are doing with their offhand can lead to a significant reduction in injuries.

After the follow-up evaluation has been completed, most workplaces will choose to embark on a schedule that incorporates an annual (internal) occupational health and safety audit.

SETTING NEW GOALS

If the evaluation process reveals that most of the goals identified at the planning stage have been achieved, management may wish to consider identifying new goals to guide the process of continuous improvement. The nature of these goals will depend on the calibre of the company's performance. For a safety "straggler" that entered the process with a high rate of injuries, a history of paying penalties under WSIB's incentive programs, poor morale, and serious gaps in compliance, a basic degree of compliance and stability—in terms of the company's safety record—may have been the initial goal. Once that goal is achieved, the company may decide that the next challenge will be to maintain its newly achieved degree of safety "competence" for the next calendar year, without relapsing into old patterns.

For a company that entered the process with a basic level of performance and compliance in place, an appropriate new goal may be to achieve a specified reduction (in terms of percentages, or person-days lost) in its rate of injuries over the next year.

For a company that already performs above the industry average, an appropriate goal might be to build a reputation for itself as a industry leader. This may include, for example, participating in initiatives to develop standards and best practices for the industry as a whole.

Health and safety experts emphasize the need to view health and safety program renewal as a continuous process. Safety innovations are constantly emerging, and a static OHS program will quickly fall behind. Work processes change, too, and along with change—even positive change—comes unfamiliarity, which can breed risks. In order to achieve excellent safety performance year after year, employers must commit sustained attention to the process of safety system renewal. Attention to safety allows for the prevention of accidents, and a little effort devoted to prevention has the potential to save time, money, and even lives in the long term.

REVIEW QUESTIONS

1. List the basic steps in the OHS system development or renewal process.
2. Why is it important for senior management in an organization to be responsible for setting OHS goals?

3. List at least two benefits of adopting an occupational health and safety "system" in the workplace.

4. Which aspects of the OHS status quo should an organization document before planning for change?

5. Why might an organization opt to seek the assistance of outside consultants in carrying out aspects of OHS system development or renewal?

6. List at least two kinds of "optional" health promotion programs that employers can provide in the workplace or to workers.

7. Why should an employer conduct a follow-up assessment of the OHS after the process of development or renewal is complete?

8. What is the last step in the process of OHS development or renewal?

CHAPTER 9
Steps to Compliance II (Post-Incident)

CHAPTER OBJECTIVES

After completing this chapter, you should be able to:

- List at least four kinds of workplace incidents that are relevant to occupational health and safety.
- Explain the importance of addressing internal and external complaints.
- Know how to respond to work refusals.
- Describe the reasons for tracking near-miss and no-injuries accidents.
- Understand your accident reporting responsibilities under the OHSA and WSIA.
- Define the terms "critical injury" and "occupational illness."
- Understand the potential impact of traumatic events in the workplace on workers.
- Explain the importance of post-incident investigations.

INTRODUCTION

The previous chapter provided an outline of the steps that an organization might consider while seeking to implement an effective OHSMS and safety program in a workplace. Unfortunately, even in the most safety-conscious and safety-compliant workplaces, accidents occasionally do happen.

When they do, the employer's reaction sends important messages to a number of different stakeholders: the injured worker, other workers at the workplace, the WSIB, the MOL, and the general public.

Timely, empathetic, and appropriate responses to workplace accidents reinforce the culture of safety in a workplace and can limit the fallout from an accident in a number of ways.

For example, the New Zealand *Aftermath* study mentioned earlier found that the employer's actions subsequent to a workplace accident had the potential to improve rehabilitation for injured workers. As well, the more involved an employer was in the wake of an accident, the better—for both the worker and the company: "[t]he most successful outcomes for both the worker and workplace were when the employer took an active role with the worker and appropriate responsibility for what occurred" (Adams et al., 2002, p. 15).

Appropriate employer responses can also reduce the likelihood of a MOL investigation, and can influence the quantum of fines in the event of a prosecution. Proper handling of incidents can also support a cooperative relationship between the employer and the WSIB.

The employer's reaction to an accident will be observed closely by other workers, who will draw conclusions from the employer's actions. Where an employer pays lip service to the notion of safety as "everyone's concern" but responds to accidents by placing the blame on the injured worker, employees will tend to give more weight to the employer's actions than to its policies in forming their opinion of the workplace culture.

An employer that fails to investigate an incident properly loses an important opportunity for prevention. While the term "accident" implies that injuries happen at random, in most cases it is possible to identify and address contributing factors. In this way, the employer may be able to prevent similar accidents from happening in the future.

Finally, while workplace accidents have the potential to negatively affect a business's public image, so does a business's response. Where the employer is perceived to be genuinely concerned and supportive of the injured worker's recovery, the public is likely to attribute the accident to misfortune rather than to mismanagement.

This chapter introduces the steps that an employer must take in the days following an incident. Chapter 10 will explain the employer's role and appropriate strategies in the event of a MOL investigation and/or prosecution; it will touch on the appeals process for orders made under either the OHSA or the WSIA; and it will discuss other possible sanctions in the wake of a workplace incident (for example, charges under environmental or human rights laws, and civil lawsuits).

WHAT IS AN INCIDENT?

Because the WSIA is designed to compensate workers for losses suffered due to a workplace injury or illness, accidents that do not cause injury or that cause minor injuries that require only first aid and do not result in lost time need not be reported to the WSIB.

However, from the perspective of a responsible employer committed to continual improvement of its safety program, the definition of an "accident" should be much broader. Because keeping records of *all* safety events—even those that do not lead to lost-time injuries—can assist in the prevention of future injuries, employers are encouraged to investigate, record, and follow up on *incidents*, not just accidents, and to use a broad definition for events that qualify.

Consider the following list of possible events. Some are reportable to the WSIB and/or the MOL, while others are not. Think carefully about the latter. How might they, if repeated under only slightly different circumstances, result in claimable injuries?

1. An employee is killed when scaffolding collapses.

2. An employee loses two fingers after her hand is pulled between two rollers on a piece of machinery.

3. An employee breaks her arm after tripping on an electrical cord.

4. A beauty salon employee complains of nausea and dizziness by mid-afternoon most afternoons. That same week, the salon owner reads, in a trade publication, that certain less expensive manicure products for nail airbrushing have been associated with similar symptoms. The salon has recently changed suppliers for its manicure supplies.

5. An employee suffers temporary stinging and burning to an eye, relieved by flushing with water, after she puts a container of dishwashing detergent down on an uneven surface and it splashes upward out of the bottle.

6. An employee complains of an acrid chemical smell on a manufacturing floor, and an investigation reveals that an overheating machine has caused a protective floor mat to melt.

7. The owner of a neighbouring business complains that runoff from the employer's textile plant is seeping onto his property and is killing his grass.

8. After receiving a negative performance review, a worker returns to his cubicle and slams his chair into his desk, knocking the computer monitor on an adjoining desk onto the floor. No one is injured.

9. High winds overnight cause an awning to be torn from the side of a building and to cartwheel across a parking lot, damaging a security guard's car.

10. Young people gain access to a construction site during the night and displace items. They bump into a wooden access ramp and replace it incorrectly. The next day, a worker notices the ramp's instability before anyone has a chance to be hurt.

11. A lumberyard owner fires a worker for theft. Out of respect for the dismissed worker's privacy, the reason for the dismissal is not communicated to the other workers. The employer hires a female employee to fill the position; she is the first female worker in the business's history. The other workers resent her because she has displaced a popular colleague for reasons unclear to them, and because they are worried about how a female co-worker will alter the culture at the workplace. To encourage her to resign, they repeatedly delegate job tasks requiring very heavy lifting to her. She complains to management.

12. A worker arrives in a senior manager's office to complain about excessive noise in an area of the plant. The worker says that he leaves work with a headache most days. He explains that he has already complained three times to his direct supervisor, but that the supervisor has told him that he is exaggerating. The supervisor has never reported the claim to his own superior, and has never suggested that the worker contact a JHSC member.

Of these scenarios, only scenarios 1 to 4 are likely to trigger a claim under the WSIA, and only scenarios 1 and 3 require the employer to make a critical injury report to the MOL. However, each of the scenarios involved the *potential*—which, luckily, remained unrealized—for a significant injury or illness and a WSIA claim. Two of the scenarios, 7 and 11, also raised the possibility of consequences under other legislation (environmental laws and human rights laws, respectively).

For an employer who is truly committed to improving safety in the workplace, all of these events should be treated as safety incidents and should receive proper documentation and follow-up. After a discussion of the rationale for tracking environmental and human rights incidents within the safety program (as well as within the employer's environmental and human rights management programs), this chapter will describe the basic steps that should be taken in the wake of various kinds of incidents.

WHY TRACK ENVIRONMENTAL AND HUMAN RIGHTS INCIDENTS?

As you have learned, incidents that raise environmental or human rights concerns often have safety implications as well.

Environmental "accidents" are tracked and prosecuted by the government because they have the potential to damage our natural environment, to threaten plants and animals, or to contaminate lands and waterways. Many events that have this potential also have the potential to harm human health.

A typical environmental incident involves the "escape" of a harmful substance into the natural environment. The fact that an employer has allowed this escape suggests that harmful substances are not being properly managed at the workplace, and that employees may be at risk for harmful exposures. Where an environmental incident has occurred, personnel with responsibilities for worker safety should analyze the circumstances and assess policies and procedures with respect to the handling of harmful substances, so that any problems can be identified.

The connection between safety and human rights is more subtle. A typical human rights incident involves a complaint by a worker (or potential hire) about unequal treatment based on a characteristic recognized as a ground for discrimination, such as age, gender, race/colour/ethnicity/origin/culture, religion/creed, disability, pregnancy, sexual orientation, or family status. While many human rights problems have no safety implications, any time harassment escalates to violence (which is broadly defined to include anger; see the definition of "violence" in chapter 6), the victim's safety is placed in jeopardy.

Another situation in which discrimination is dangerous occurs when workers refuse to work with the harassment victim in the manner required by the job. An example might be where a harassment victim is forced to perform a work task alone that is designed for two people working together.

Finally, where discrimination is the basis for improper delegation of work (as you saw above, in the lumberyard example), workers who are the targets of discrimination may be forced by colleagues to do harder, heavier, dirtier, riskier, or more uncomfortable work. The safety implications of this are clear.

Environmental problems and human rights problems can easily become safety problems. For this reason, the safety system should include the monitoring, reporting, and follow-up of these incidents.

HARD LESSONS
What if Law Is Not Enough?

In 1988, frustrated from a struggle with unrelenting racism in his workplace, correctional worker Michael McKinnon brought the first of a long line of human rights complaints against the Ontario Ministry of Correctional Services.

McKinnon, who is of aboriginal descent, had endured ongoing harassment, intimidation, and discrimination at the hands of both co-workers and inmates. He witnessed similar discrimination against other employees, and decided to take action in the form of a human rights complaint.

Ten years later, after the appointment by the Human Rights Tribunal of Ontario (HRTO) of a special Board of Inquiry to look into the matter, the HRTO acknowledged that the Toronto Metro East Detention Centre was afflicted with serious, systemic racism and was a "poisoned work environment." The HRTO issued a long list of orders designed to remedy the problems, both in the specific detention centre from which the complaint originated and in the Ministry as a whole. Examples of the orders issued included:

- compensating the complainant for acts of discrimination and reprisal;
- promoting the complainant (who had been passed over for promotion on racist grounds);
- relocating workers who had been involved in racist acts;
- reading out the orders made in public at the workplace and attaching a copy to employee pay stubs; and
- instituting a training program designed to improve cultural sensitivity and to prevent discrimination on the basis of race.

The facility and the Ministry complied with the orders, but unfortunately, the situation at the Metro East Detention Centre did not change. Harassment and discrimination on the basis of race continued, and the poisoned environment remained. McKinnon made new complaints to the HRTO, and further orders were issued. In subsequent litigation, the HRTO ruled that the Ministry failed to make a good-faith effort to comply with the new orders. In a 2005 ruling the HRTO explained:

> Despite protestations to the contrary, it seems to me that the Ministry is not truly focussed on that end as a positive institutional good to be desired by all affected thereby, from the top down, but views the orders as an unavoidable imposition that it wants both to limit and to have over with as soon as possible. Although it appears that the Ministry is still seeking "closure to the process"—an apparent expression of lament, rather than of hopeful anticipation, it seems reluctant to pass up any perceived opportunity to attempt to restrict the ambit of reform. (*McKinnon v. Ontario (Correctional Services)*, 2005 HRTO 15)

In January 2007—nearly 19 years after the original complaint—Michael McKinnon joined with Leah Casselman, president of OPSEU, to call upon Ontario Premier

> Dalton McGuinty to intervene. McKinnon and Casselman told the premier that despite nearly two decades of litigation and court rulings in McKinnon's favour, nothing had changed, and that there were at least 20 other unresolved human rights complaints pending against the jail.
>
> McKinnon has been off work for the past three years because the HRTO has concluded that the environment in the jail is such that it is unsafe for him to work there.
>
> The McKinnon case is unsettling because it reminds us that there are limits to the efficacy of legal solutions to workplace problems. Michael McKinnon mustered the courage to take on the system, and he won—more than once. And yet, conditions in the prison did not change. Despite the best efforts of the HRTO, attitudes in the complainant's workplace proved stubbornly resistant. The most recent round of litigation resulted in the HRTO making 34 new orders in July 2007; whether these orders will be complied with in good faith remains to be seen.
>
> This case underlines the need for a multidisciplinary approach to workplace problems, whether they be human rights problems, safety problems, or—as in this case—a little of both. (For more information, see *McKinnon v. Ontario (Correctional Services)*, 2007 HRTO 4.)

BASIC STEPS FOLLOWING AN INCIDENT

The following sections will outline the first steps to be taken after each of the following types of incidents:

- complaints about hazards/conditions
- work refusals/stoppages
- accidents without injuries (property damage, environmental accidents, human rights problems, and injury near-misses)
- first aid/no-lost-time injuries
- lost-time injuries
- critical injuries and fatalities
- occupational illnesses
- traumatic events

Some of these events involve reporting requirements, WSIA claim support obligations, obligations with respect to JHSCs, and obligations with respect to a union (if any) in the workplace. These obligations will be "flagged" in the "first steps" discussions, with a fuller discussion to follow below under separate headings; a general discussion about post-incident investigations appears at the end of the chapter.

COMPLAINTS ABOUT HAZARDS/CONDITIONS

There are two general classes of safety-related complaints that can come to an employer's attention. Internal complaints come directly from a worker or supervisor or are communicated via the Health and Safety Representative or JHSC; external complaints are communicated to the employer by a source outside of the workplace, for example, the WSIB or MOL.

Internal Complaints

Internal safety-related feedback and complaints should be *encouraged*, rather than discouraged, by employers. Where the workplace environment is one in which workers feel comfortable raising safety concerns as soon as they develop, managers can take steps to address these concerns before they trigger work stoppages or even accidents. Front-line supervisors in particular should be instructed to welcome feedback and to communicate it promptly to more senior management.

In general, the appropriate first step for a worker who has a safety concern is to speak with his immediate supervisor. The supervisor should then either take steps to solve the problem or refer the problem to a more senior manager for resolution. Only where the worker is not happy with the actions taken by the supervisor or other employer representatives will he need to take further action, by communicating his concern to the Health and Safety Representative or a JHSC member.

When a complaint has been made to the Health and Safety Representative or JHSC, the representative or JHSC is required to investigate the complaint and to make recommendations to the employer, the worker, and any other appropriate parties (for example, a union).

Health and Safety Representative or JHSC recommendations require a written response, by the employer, within 21 days. The response must contain either a timetable for the implementation of an employer solution to the problem or, where the employer elects to take no action, a statement outlining the reasons for not taking action. If the Health and Safety Representative or JHSC is not satisfied with the employer's response, the next step is usually a phone call to a MOL inspector to request an inspection.

It is more efficient, of course, to prevent safety issues from getting this far by encouraging open dialogue about safety. Worker feedback will only flow freely, however, in an environment in which workers witness timely and appropriate reactions to such feedback. It is not good enough for management to simply thank workers for their suggestions; workers must also see that the employer is willing to take steps to implement their suggestions and resolve problems.

Where a worker's suggestion has led to a safety improvement, it is often appropriate—and good business—to recognize the worker's contribution in a public way. Among other things, this will provide other workers with an example to follow. Recognition need not be financial or excessive; posting a photo of the worker and a memo about her role in the safety improvement on the staff room bulletin board can be sufficient. In larger offices, recognizing "safety heroes/leaders/innovators" on a regular basis (perhaps by choosing one leader per quarter and profiling him or her in a newsletter) is an excellent way to reinforce workers' participation in the success of a safety program.

In a workplace where safety is an open dialogue, feedback should go both ways. As noted earlier, a culture of safety demands that supervisors "challenge" unsafe worker behaviours whenever these are observed. Knowing how to challenge a worker effectively—with authority, but also supportively, and with tact—should be a skill that is taught to every supervisor.

Finally, all internal complaints (regardless of the source, or whether they come from or are made to a worker) should be recorded in an appropriate file. Action taken in response to a complaint should also be recorded. Issues that seem trivial at the time that they arise may eventually prove to be important for some reason somewhere down the line, and good records provide backup when memory falters.

External Complaints

If proper procedures are followed and if the workplace nurtures a safety culture characterized by open dialogue, safety issues should never come as a "surprise" to employers via complaints from outside sources—for example, a visit by a MOL inspector who has received an anonymous phone call. This situation is a clear red flag for problems with the workplace safety culture.

While the MOL is entitled to conduct its inspections at random, the realities of limited resources generally motivate inspectors to visit workplaces where there is reason to believe there is a problem. There is no requirement that an inspector give notice to the employer before arriving at the workplace.

Where an employer does receive notice before an inspector's visit, preparation for the inspection should include:

- arranging for the Health and Safety Representative or a "worker member" (preferably certified) of the JHSC to accompany the inspector. If neither is available, a union representative or a person chosen by the workers to represent them may accompany the inspector instead;
- arranging for the availability of employer records for inspection (for example, maintenance records, incident records, and minutes of JHSC meetings);
- arranging for the availability of workplace plans;
- if the employer suspects that the inspection relates to a particular safety concern or problem, requesting a briefing about the situation from appropriate parties (supervisors, the JHSC, or Health and Safety Representative) so that the employer is not "in the dark" about the purpose of the inspection.

Any employee who is chosen to accompany the MOL inspector is entitled to payment for the time spent doing so.

If the MOL inspector identifies any problems at the workplace, she may issue orders (oral or written) requiring the employer to remedy them. In some cases, inspectors may post notices of their orders in the workplace for workers to see (for example, where an order requires a machine to be repaired before it can be used again). An employer is typically required to comply with MOL orders within a timeframe specified by the inspector, and in some cases may be required to file a compliance plan with the MOL.

Once the employer has complied with a MOL order, it must forward (within 3 days of complying) a "notice of compliance" to the MOL. The employer must also post the order and its notice of compliance in the workplace, in a place where the posting is likely to be seen by workers.

Where an employer learns that a MOL inspection was launched in response to a complaint, by an employee, to the MOL, the employer is prohibited from making reprisals against the employee (OHSA, section 50); the prohibition against reprisals was discussed in chapter 3.

Appeals, by the employer, of MOL orders will be touched on in chapter 10.

In addition to the MOL, another source of external "complaints" can be the WSIB. Examples of situations in which employers may find themselves in conflict with the WSIB include:

- the employer's failure to report reportable accidents;

- the employer's failure to file necessary documentation in support of a worker's claim;

- suspicion, on the part of the WSIB, that the employer is misrepresenting its status (for example, the size of its payroll) in an attempt to evade premiums;

- suspicion of employer fraud by the WSIB (for example, where the WSIB suspects that an employer has colluded with a worker's filing of a fraudulent claim);

- the employer's failure to support the worker in his early and safe return to work, including failure to properly accommodate the worker; and

- the employer's obstruction of or failure to comply with a WSIB investigation into any matter within its powers.

Where an employer does not resolve a conflict with the WSIB, the WSIB is sometimes entitled to levy fines against the employer. See, for example, section 41(13) of the WSIA, which allows the WSIB to levy a fine against an employer who fails to support the worker's return to work, and then pay the money to the worker as if the worker were entitled to a loss of earnings benefit under section 43. Where an employer files premium information or pays premiums late, the WSIB can charge interest. In some cases—for example, in the case of a suspected fraud—the WSIB can commence a prosecution against the employer for an offence under the WSIA.

Finally, an external "complaint" may come in the form of a criminal charge (for example, for serious negligence causing injury; see section 22.1 of the *Criminal Code*, which was discussed in chapter 2 in the context of the Westray mine tragedy) or in the form of an investigation or prosecution under environmental or human rights legislation. In each of these cases, procedures for responding to these charges or complaints are outlined in the legislation under which the investigations are launched.

WORK REFUSALS OR STOPPAGES

Sections 43 to 49 of the OHSA establish the right of workers to refuse unsafe work or to stop work that they consider unsafe.

When a worker has decided that the work he has been asked to do or that he has been doing is unsafe, he is entitled to stop work immediately. A worker who exercises his right to stop work must report the stoppage to the employer immediately (this includes a supervisor) and explain his reason(s) for stopping work. The bases recognized under the OHSA for a work stoppage or refusal are as follows:

- A machine, piece of equipment, or tool that the worker is using or is told to use is likely to endanger him or another worker (section 43(3)(a)).

- The physical condition of the workplace or workstation is likely to endanger the worker (section 43(3)(b)).
- A machine, piece of equipment, or tool that the worker has been asked to use or the physical condition of the workplace contravenes the OHSA or the regulations, and is likely to endanger the worker or another worker (section 43(3)(c)).

The worker is required to stand by and be present while the employer conducts an immediate "investigation" of the conditions that led to the work refusal or stoppage. The employer (or its agent, for example, a senior manager) must be assisted in the investigation by a JHSC member (preferably a certified member) or, if there is no JHSC, the Health and Safety Representative. If there is no JHSC and the Health and Safety Representative is not available, the employer must be assisted by a worker who has been chosen by the workers or the union to represent them for the purpose of the investigation.

If the immediate ("first stage") investigation results in a resolution of the problem (for example, if the worker had been asked to lift something on his own that required two lifters, and the investigation led to the summoning of a second lifter), nothing more will happen, though the employer should make a record of the incident.

If the investigation does *not* resolve the worker's concerns, the worker is entitled to continue to refuse work, and the situation becomes a "second stage" investigation. At this point, the employer is required to notify a MOL inspector, who is required to investigate the problem.

The MOL inspector can either attend at the workplace as soon as possible, or can "investigate" by speaking with the employer's representative, the worker, the worker's representative (JHSC member, Health and Safety Representative, or other designated employee, or designated worker investigator) and a union representative, if there is a union.

After the worker has had an opportunity to speak with the MOL inspector and/or observe the inspection, but before the problem is resolved, he is expected to remain at the workplace. During this time, he can either wait a safe distance from the workspace or do alternate safe work assigned by the employer. In cases where the inspection is expected to be lengthy and there is nothing else for the worker to do, he will be allowed to go home, with no loss of pay.

Pending the results of the investigation, the employer is entitled to ask a different worker to do the work that the first worker refused. The replacement worker must be told about the work refusal and the reasons given by the first worker; if she perceives a risk, she too is entitled to refuse the work.

The MOL inspector must deliver the results of her inspection, in writing, to all the parties (the employer, worker, and representatives). If the inspector has found that the stoppage or refusal was founded, no worker is permitted to do the work until the employer has made the necessary changes. If the inspector has found that the stoppage or refusal was not warranted, the refusing worker is expected to return to work immediately.

A worker can never be disciplined or penalized by the employer in any way for stopping or refusing work as prescribed by the OHSA. If the employer does penalize a worker for doing so, the worker can begin a **grievance** under his collective agreement (if he has a union) or can file a complaint with the OLRB. Both of these actions are touched on in chapter 10.

grievance
a formal allegation that something is unjust, inequitable, or illegal in unionized workplaces

ACCIDENTS WITHOUT INJURIES

Accidents that do not cause injury are often called "near misses" or "near hits." Examples of such incidents might include:

- a structure collapses while nobody is on, in, or near it;
- an employee catches his sleeve on a moving part but is able to stop the machine before injuring his arm;
- a health care employee discovers a needle in the regular trash while bagging it, but is not pricked;
- a container containing a caustic liquid is dropped, but all nearby employees step aside in time to avoid being splashed.

While it may be tempting to forget these potentially frightening events as quickly as possible, ignoring or concealing near misses eliminates key opportunities for identifying workplace risks.

Safety-conscious employers should encourage the reporting of near misses, and should have a procedure in place both for making the reports and for investigating those incidents deemed to be most serious. The purpose of near-miss investigations must not be to determine fault, but rather to identify the conditions that led to the accident so that similar accidents can be prevented in future. Employers should encourage a workplace culture in which near-miss reports are treated as a learning opportunity, not as grounds for warnings or discipline. Some ideas for turning near misses into opportunities to improve safety include:

- encouraging workers to come up with solutions to safety gaps that have led to near misses, with awards (and/or rewards) offered for implemented solutions;
- including a discussion of recent near misses at worker team meetings; and
- after addressing internal near misses, sharing best practices with industry peers through professional associations or other forums.

Where an incident results in no injuries but causes property damage, the damage should be documented for insurance purposes or, if the incident appears to have been caused by faulty equipment, to support attempts to obtain replacement equipment or repairs.

Where there has been an exposure incident (an incident that has caused workers to be exposed in some way to hazardous materials) but where there are no lost time injuries or ongoing illnesses, the employer can voluntarily fill out the WSIB's Employer Exposure Incident form (Form 3959A; available on the WSIB's website) to document the exposure. The form is designed to allow employers to record information about the exposure, which can be referenced later should an exposed worker become ill.

FIRST AID/NO-LOST-TIME INJURIES

Accidents that result in minor injuries—the kind that require only first aid and do not result in lost time for the worker—need not be reported to the WSIB.

However, while such incidents need not be reported, they must still be properly attended to. The employer is required to have a first aid kit on site. In workplaces where there are known injury risks, it is prudent to designate certain employees as first-aiders, and to arrange for first aid training for these workers.

Appropriate first aid supplies should be kept on hand. Guidelines for the contents of a basic first aid kit are available from many sources, including St. John Ambulance (www.sja.ca), public health authorities, and the CCOHS. Kits should include supplies designed to protect first-aiders, for example, latex gloves, one-way valve masks for CPR, and disinfecting wipes. All workers should be made aware of the location of the first aid kit(s), and a program should be in place for the review of the kit's contents and replacement of depleted supplies.

When first aid is provided to a worker, details of the treatment and the identity of the provider should be recorded for future reference. The worker should be asked whether he or she intends to obtain follow-up care from a health care practitioner; if the worker does intend to do so, the employer may need to make a report to the WSIB, and WSIB Form 0156C (Treatment Memorandum) may need to be completed to permit the health care professional to submit his or her account to the WSIB. If, after receiving first aid, the worker wishes to go to the hospital or to visit a health care practitioner, the employer must arrange and pay for the worker's transportation to the hospital. (Once a worker elects to go to a health care provider for treatment, the injury is no longer a first-aid-only incident.)

Once first aid has been provided, the worker should be permitted to decide whether or not he or she is well enough to remain at work. The WSIA requires that the employer pay the worker's full wages for the day of the accident, regardless of whether the worker works the full day; leaving early on the day of the accident does not count as lost time for the purpose of reporting duties under the WSIA.

Once the worker's injury has been fully attended to, the incident should be documented and investigated in an effort to prevent future similar incidents.

LOST-TIME/MEDICAL-ATTENTION INJURIES

The next class of incidents are those that require treatment or recuperation beyond simple first aid but do not meet the definition of "critical injury" as defined by Regulation 834 made under the OHSA.

Immediate Considerations

When an accident happens, work in the vicinity of the accident should be stopped immediately and the area left undisturbed. This will allow the employer to investigate the causes of the accident. First, however, the injured party/parties must be attended to.

When it becomes apparent that an injury is serious enough to require the attention of a health care practitioner and/or time off work for recovery, the worker will almost always leave the workplace. He or she will either go directly to a health care practitioner (the employer is required to arrange and pay for transportation) or will go home (it is advisable for the employer to arrange and pay for transportation).

If it makes sense to do so from a practical standpoint, the employer should provide first aid through a designated first-aider while awaiting transport. The aim of this care should be to minimize aggravation of the worker's condition. First aid steps may include:

- efforts to stop or slow bleeding;

- immobilization of injured limbs;

- rinsing off of hazardous substances;

- administering physician-directed treatments designed to minimize poisoning; and

- making the injured person comfortable (for example, ensuring that the person is warm can reduce the risk of shock).

Employers should be prepared for serious injuries by knowing what kinds of health care services are available in the area (the employer should know, for example, where the nearest emergency room is, and what number to call for poison control).

When a worker leaves work to seek medical treatment or to go home to recuperate, the worker is entitled to wages for the entire shift for which she was scheduled to work on that day.

Once the worker is safely on his way to treatment, the JHSC or Health and Safety Representative should be contacted, if they have not been already. (In workplaces where there is a union with representatives on site, it is good practice to advise the union as well.) An investigative group comprising an employer representative (or representatives) designated to deal with investigations, the Health and Safety Representative or a JHSC member (preferably a worker member), and a union representative should inspect the scene of the accident. The goal of this inspection should be to determine, if possible, the cause(s) of the accident, and to document any unusual conditions (for example, a slippery floor, a missing or broken piece of equipment, obstacles or debris, and so on). The investigative team should speak with any workers who were present at the time of the accident to obtain eyewitness accounts.

The results of the investigation should be recorded in writing, with copies made for all of the parties. These results may be needed for the purpose of fulfilling reporting obligations or for assisting with a WSIB claim.

Reporting Obligations Generally

Whenever a worker requires medical attention beyond first aid for a workplace accident, the incident must be reported to the following parties:

- an inspector (and in some cases, a Director) from the MOL (OHSA, section 52(1));

- the WSIB (using Form 7, as prescribed by section 21 of the WSIA);

- the Health and Safety Representative or JHSC (OHSA, section 52(1)); and

- the worker's union, if any (OHSA, section 52(1)).

Reporting Obligations Under the OHSA

Sections 51 to 53 form part VII of the OHSA. They describe the procedures employers and constructors must follow in the aftermath of a workplace accident or the diagnosis of an occupational illness. Section 52(1) deals specifically with non-fatal, non-critical accidents:

> If a person is disabled from performing his or her usual work or requires medical attention because of an accident, explosion or fire at a workplace, but no person dies or is critically injured because of that occurrence, the employer shall, within four days of the occurrence, give written notice of the occurrence containing the prescribed information and particulars to the following:
>
> 1. The committee, the health and safety representative and the trade union, if any.
>
> 2. The Director, if an inspector requires notification of the Director.

As you can see, the notice prescribed under the OHSA must be given within four days of the employer's becoming aware of the accident.

Reporting Obligations Under the WSIA

While the employer's reporting obligation under the OHSA is chiefly for government record-keeping purposes, the report that the employer must make to the WSIB has two key practical functions: it provides information in support of the worker's claim for compensation, and it provides premium-setting information to the WSIB.

Under the WSIA, the employer must report non-critical injuries not only in circumstances where the worker seeks medical treatment, but also where the worker does not need treatment but does miss regularly scheduled work time (beyond the day of the accident) or requires modified work.

For the purpose of the WSIB report, medical treatment includes treatment by:

- a physician;
- a nurse;
- a chiropractor;
- a physiotherapist; or
- a dentist.

Also, if the accident causes damage to a worker's glasses, dentures, hearing aid, prosthetics, and so on, the employer is required to make a report to the WSIB, even if the worker herself does not suffer any injury.

Missing work includes situations in which a worker must cut back his usual working hours (for example, because he cannot sit/stand/read for extended periods) and situations in which the injury requires that the worker return to work in a different capacity for reduced pay.

Where a worker can do only modified work for a period of more than seven days even though her pay remains the same, the employer must file an injury report.

The section that imposes the WSIB reporting requirement is section 21(1) of the WSIA:

An employer shall notify the Board within three days after learning of an accident to a worker employed by him, her or it if the accident necessitates health care or results in the worker not being able to earn full wages.

A copy of this notice must also be given to the worker, on the same day. Failure to make the report to the WSIB or to the worker, reporting late, or submitting an incomplete report can lead to fines (usually $250 for each infraction, which means that an employer who files a report that is late *and* incomplete will be fined $500).

The WSIB requires that this report be made on the most up-to-date version of its form for the purpose. At the time this chapter was written, the applicable form was Form 0007A. The form, along with a useful guide for its completion, is available from the WSIB, and the current version is included in appendix I.

The WSIB's Form 7 Reference Guide for Employers provides step-by step guidelines on how the form should be completed, and so those instructions will not be repeated here. Some points of particular note, however, are the following:

- The form has a box for the claim number. The WSIB will provide this once a claim has been initiated. It should be left blank if no claim yet exists.

- In the section for describing the worker's job, fill in the job the worker was doing *at the time of the accident*. For example, if the worker is employed in a fast-food restaurant as a cashier but was deep-frying chicken at the time of the accident, the occupation should be listed as "food preparation" or similar—not "cashier."

- Firm/account numbers are assigned to the employer by the WSIB for the purpose of charging premiums. An employer may have a single firm number (for example, if all the employees fall into a single classification), or several account numbers. The number that corresponds to the particular worker's classification should be given.

- In the description of Business Activity box, the employer can use the description that appears on the top right corner of its premium remittance statement (for example, "bicycle repair shop").

- The employer must complete a description of what the worker was doing at the time of the accident or injury. The WSIB requests that, if the employer has completed a physical demands analysis of the work function, a copy of that analysis be attached to Form 0007A as an appendix. (The WSIB's form 2830A, available on its website, can be used to guide the employer through a physical demands analysis).

- There is a space on the form for describing the employer's awareness, if any, of "any prior similar or related problem, injury or condition." This is an important question. In some cases, an accident leads to the aggravation of an existing health problem or injury, and the existence of the prior injury can affect the way in which the claim is treated. If the employer knows of a pre-existing health problem (for example, a worker who falls and strains his back has had a history of back problems stemming from a previous career), the details known to the employer should be provided here.

- There is a box that the employer can check to indicate that she has appended a summary of any "concerns about this claim." There are no specific guidelines provided for what might be included in this summary, though the term "concerns" is fairly broad and general. If, for example, the employer suspects that the claim is fraudulent (for example, that the worker is greatly exaggerating the extent of her injury or illness), this would be the place to describe those suspicions. However, an employer should reflect very carefully before reporting any suspicions that are not grounded in objective evidence.

- There is a section for reporting whether the worker "received any health care." This does *not* include first aid. (For an explanation of what the WSIB considers first aid, read page 4 of the WSIB's *Reference Guide for Employers*.)

- In the section for reporting lost work time, missing work (including a partial shift) to attend a health care appointment counts as missed work time.

- Form 7 includes a detailed section dealing with return-to-work issues. In order to complete this section fully, employers may need to do two things:
  1. obtain/complete a Functional Abilities form to assist in planning for modified work, if warranted; and
  2. prepare a (written) offer of modified work for the worker's consideration. If the worker declines this offer, the written offer will need to be appended to the Form 7.

- Form 7 requires a detailed calculation of the worker's earnings, at least where there has been lost work time (the section can be completed in more general terms where there is no lost time). The calculations require that the employer take into account such things as wages, salary, commission, vacation pay, and bonuses. Where there are special circumstances that make it difficult to describe the work contract in the terms used by the form, the employer is encouraged to contact a WSIB adjudicator to discuss the issue(s).

Once the Form 7 is complete, the worker can be asked to sign it, expressing his agreement that the information is accurate and giving his consent to the disclosure of any information necessary for conducting a functional abilities analysis. If the worker declines to sign the Form 7, he or she can provide consent to disclosure using Form 1492A, the Worker's Claim/Consent form.

In order to initiate a claim, the worker has his own form—Form 6—to complete. It is good practice for the employer to advise the worker of this responsibility, and of the six-month period within which the worker (or his survivors) must file.

FACTS AND FIGURES

Other Post-Injury Reports

Besides the employer's Form 7 and the worker's Form 6, there may be other paperwork to complete in the wake of an accident. For example, where a worker seeks to have treatment reimbursed by the WSIB, the employer will need to complete Form 0156C, Treatment Memorandum. If the worker has complained of hearing loss related to working conditions, the employer may need to complete Form 0137A, Employer's Report Occupational Noise Induced Hearing Loss.

CRITICAL INJURIES AND FATALITIES

Section 51 of the OHSA deals with the most serious accidents: those that result in a worker's death or critical injury.

FACTS AND FIGURES

What Is a Critical Injury?

Section 51 of the OHSA requires employers to report any accident resulting in death or a critical injury immediately to the MOL. But what exactly constitutes a critical injury?

The term "critical injury" is defined in its own regulation—RRO 1990, reg. 834—as follows:

"critically injured" means an injury of a serious nature that,

(a) places life in jeopardy,

(b) produces unconsciousness,

(c) results in substantial loss of blood,

(d) involves the fracture of a leg or arm but not a finger or toe,

(e) involves the amputation of a leg, arm, hand or foot but not a finger or toe,

(f) consists of burns to a major portion of the body, or

(g) causes the loss of sight in an eye.

Immediately after a serious accident, the employer (and constructor, if any) must immediately notify the following people or groups:

- an inspector at the MOL;

- the Health and Safety Representative or JHSC; and

- the worker's labour union, if any.

This notification should be done in person or by telephone, *not* by mail/courier, email, or fax.

Although the provision does not specifically say so, the employer must also arrange for first aid and the immediate transport to hospital of any critically injured worker. In addition, the employer must take steps to preserve the scene of the accident for investigative purposes. This means stopping work in the immediate area and ensuring that no one will "interfere with, disturb, destroy, alter or carry away any wreckage, article or thing at the scene of or connected with the occurrence until permission to do so has been given by an inspector."

There are three narrow exceptions to the rule against disturbing the wreckage left in the wake of an accident. An employer may move something if doing so is necessary to:

- preserve human life or alleviate human suffering;

- maintain "an essential public utility service or a public transportation system"; or

- prevent unnecessary damage to property or equipment in the vicinity (for example, if the injury occurred due to a fire, and something combustible, such as a hallway runner, was still burning, the employer could remove the burning object to the parking lot to avoid allowing the fire to spread).

Within 48 hours of a serious accident, the employer is required to file an occurrence report describing the accident to a MOL Director. The form of this report is provided in the Industrial Establishments regulation (section 5(1)), which reads:

The written report required by section 51 of the Act shall include,

(a) the name and address of the constructor and the employer;
(b) the nature and the circumstances of the occurrence and the bodily injury sustained;
(c) a description of the machinery or equipment involved;
(d) the time and place of the occurrence;
(e) the name and address of the person who was killed or critically injured;
(f) the names and addresses of all witnesses to the occurrence; and
(g) the name and address of the physician or surgeon, if any, by whom the person was or is being attended for the injury.

Section 5(2) explains the requirements for the reporting of a less serious accident:

(2) For the purposes of section 52 of the Act, notice of,

(a) an accident, explosion or fire which disables a worker from performing his or her usual work; or
(b) an occupational illness,

shall include,

(c) the name, address and type of business of the employer;
(d) the nature and the circumstances of the occurrence and the bodily injury or illness sustained;
(e) a description of the machinery or equipment involved;
(f) the time and place of the occurrence;
(g) the name and address of the person suffering the injury or illness;
(h) the names and addresses of all witnesses to the occurrence;
(i) the name and address of the physician or surgeon, if any, by whom the person was or is being attended for the injury or illness; and
(j) the steps taken to prevent a recurrence.

(3) A record of an accident, explosion or fire causing injury requiring medical attention but not disabling a worker from performing his or her usual work shall be kept in the permanent records of the employer and include particulars of,

(a) the nature and circumstances of the occurrence and the injury sustained;
(b) the time and place of the occurrence; and
(c) the name and address of the injured person.

(4) A record kept as prescribed by subsection (3) for the inspection of an inspector shall be notice to the Director.

The reports required to be filed under sections 51 to 53 of the OHSA do *not* replace or eliminate the need to file Form 7 with the WSIB.

OCCUPATIONAL ILLNESSES

What Is an Occupational Illness?

An "occupational illness" is defined under the OHSA (section 1(1)) as:

> a condition that results from exposure in a workplace to a physical, chemical or biological agent to the extent that the normal physiological mechanisms are affected and the health of the worker is impaired thereby and includes an occupational disease for which a worker is entitled to benefits under the *Workplace Safety and Insurance Act, 1997*.

Under the WSIA (section 2(1)), the definition of "occupational disease" includes:

(a) a disease resulting from exposure to a substance relating to a particular process, trade or occupation in an industry,

(b) a disease peculiar to or characteristic of a particular industrial process, trade or occupation,

(c) a medical condition that in the opinion of the Board requires a worker to be removed either temporarily or permanently from exposure to a substance because the condition may be a precursor to an occupational disease,

(d) a disease mentioned in Schedule 3 or 4, or

(e) a disease prescribed under clause 15.1(8)(d).

As was explained in chapter 5, there are two Schedules to the WSIA (Schedule 3 and Schedule 4) that describe the more common occupational diseases and establish presumptions that make it easier for afflicted workers to establish a connection between their work activities and certain specific diseases. Section 15.1(8)(d) establishes another such presumption specific to firefighters.

In chapter 5, you learned that a disease need no longer already form part of the list of recognized occupational diseases for a worker to assert that it arose through her employment. Now, the worker is entitled to initiate a claim even if the disease is not in Schedule 3 or 4, and the WSIB is required to consider the evidence in deciding whether or not the worker's illness is an occupational disease.

What to Do When a Worker Reports an Occupational Illness

Because occupational illnesses are more likely to arise gradually (as opposed to suddenly—although some *do* arise suddenly), the issue of how to determine the onset of the claimable "injury" is less straightforward than in the case of accident injuries. According to the WSIB, the relevant date, from a claim perspective, is the date on which the worker first suffered symptoms that interfered with his ability to work. In a case where a worker has been off work sick or has had to leave work early on one or more occasions due to symptoms, that date is reasonably easy to determine.

Once the employer becomes aware of a worker's alleged occupational illness, the employer is required to make the necessary reports (see the section above that deals with lost-time injuries) under the OHSA and the WSIA.

Handling an allegation of occupational illness requires a certain degree of sensitivity. When an accident happens and someone is hurt, the causation of the injury is very clear and objective: the accident caused the injury. Where an employee develops an occupational illness, the connection between the illness and the workplace can appear less definitive, especially in cases where:

- the illness is not on Schedule 3 or 4;
- no other workers are ill and there is no history, in the workplace, of occupational illness claims;
- the symptoms are not obvious; or
- the worker who is alleging the illness suffers from health problems in the first place.

The employer is urged to remember, however, that every occupational illness in every workplace has a first case. Disease is not terribly predictable: a particular level of exposure will not produce the same results in every worker who is exposed. Conditions in the workplace (temperature, humidity, ventilation, and so on) are constantly changing, and in addition, some people are simply more susceptible to illness than others. Factors influencing susceptibility can include age, genetics, nutritional status, physical fitness, stress levels, as well as *past* exposures and illnesses. For example, a person who was once a heavy smoker but who quit several years before workplace exposure to an airway irritant may be more susceptible to occupational illness from the irritant than a person who never smoked would be. When hiring workers, employers must take them as they find them; sometimes, this may mean with vulnerabilities to certain occupational diseases.

Employers must be careful to respond in a professional manner to reports of occupational illness, and should refrain from challenging workers on the basis of being the sole, or the first, sufferer.

The question of whether or not the employee's illness turns out to be an occupational illness will be answered by the WSIB after a review of medical opinions. While awaiting this determination, the employer should conduct its own internal investigation into toxic exposures in the workplace. Material safety information for *all* of the materials handled by the ill worker should be reviewed to determine which, if any, is the most likely culprit. In some cases, it will be necessary to consider whether any materials have reacted with each other in a way that made them more dangerous. In more complicated investigations, it may be useful to obtain the services of an expert, such as an industrial hygienist, to assist with the analysis of exposure hazards.

The investigation of workplace exposures must include a review of the PPE in use in the ill employee's work area. Even if the review of exposure hazards fails to identify an area of concern, it is prudent to take steps to upgrade the PPE of other workers in the area in an effort to prevent further problems. If the investigation reveals that there *is* an unacceptable exposure risk, work in the area should be stopped until the hazard has been eliminated.

TRAUMATIC EVENTS

Certain events in a workplace have the potential to affect workers even where those workers have escaped physical harm. Examples of traumatic workplace events include:

- serious fires or explosions that require evacuation and that cause employees to fear for their safety;

- serious spills or escapes of airborne contaminants that cause symptoms in workers and that provoke an evacuation;

- bomb threats or other threats (for example, suspicious packages) that cause significant fear or panic;

- armed robberies, hostage takings, or other violent crimes, whether or not there are injuries;

- outbreaks of serious communicable diseases;

- serious interpersonal violence witnessed by workers (for example, someone pointing a gun);

- any kind of workplace fatality;

- serious injuries to co-workers (for example, crushing injuries or amputations).

The impact of these kinds of events on unharmed workers is very unpredictable, and will be determined by factors such as past history (workplace violence can have a greater impact, for example, on people who have been victims of violence previously), personality (some people are naturally more fearful than others), specific phobias (sensitivity to the sight of blood, fear of contamination/disease), learned coping mechanisms, and support from friends and/or family.

For some individuals, a traumatic event will be only temporarily upsetting. In rare cases, however, a worker may suffer ongoing upset, nightmares, anxiety, or even post traumatic stress disorder (PTSD). To prevent these possible consequences (and related absences and claims) as far as possible, the employer should be prepared to offer trauma counselling to any worker who wishes it. This counselling can be offered through an existing EAP, through a specialized consultant provider to the workplace, or through a referral to a private provider (with a promise by the employer to pay, at least for a certain number of sessions). An employer should have a plan for trauma counselling in place *before* a traumatic event happens, and should already have identified and screened providers so that there is no delay in providing the services.

FACTS AND FIGURES

Post-Traumatic Stress Disorder

Post-traumatic stress disorder (PTSD) (sometimes called post-traumatic stress *reaction* or *syndrome*) is a severe psychological response to an experience that the sufferer experienced as traumatic. Usually, a clinical diagnosis of PTSD requires that the triggering event threatened the life, health, or bodily integrity of the sufferer him or herself, and not that of another person.

Most people do not suffer PTSD after a traumatic event, but instead feel grief, anxiety, and other "normal" feelings that subside over time. A person with PTSD, however, does not cope in an expected way with his or her experiences and may suffer symptoms that include nightmares, flashbacks, emotional detachment/numbness, avoidance of certain places or situations, sleep and/or appetite disturbances, irritability, memory loss, depression, and anxiety.

PTSD is treated in a variety of ways, including through counselling/therapy; with antidepressant, antipsychotic, or other drugs; and through a treatment called eye movement desensitization and reprocessing (EMDR).

Some PTSD sufferers have sufficient control over their symptoms to maintain a normal working life. Others, however, may not be capable of working—especially in an environment that they associate closely with their triggering event(s).

INVESTIGATING INCIDENTS

Introduction

The employer's duty to investigate workplace incidents that have a health and safety impact has already been touched on, both in the section above and in previous chapters. The subject does, however, bear revisiting one more time, because effective incident investigations are key to the prevention of future accidents and exposures.

The Investigation Team

As was explained in the previous chapter, the employer should be prepared for the task of investigating workplace incidents *before* the need arises. This preparation includes identifying the individuals who will conduct the investigation. In most cases, an appropriate investigative team will consist of, at minimum:

- a representative of the employer who has sufficient authority to order the implementation of changes to procedures or equipment in response to investigation findings (typically, a senior manager); and

- the workplace's Health and Safety Representative, or a worker member of the JHSC (provided that that individual is a certified member).

If the only certified members of the JHSC are management members (not an ideal situation), then a certified managment JHSC member should attend *in addition* to the representative of management, and the Health and Safety Representative or uncertified worker JHSC member.

In the wake of more complicated incidents and/or in unionized workplaces, it may be appropriate to have additional individuals participate in the investigation. These might include:

- additional JHSC members;

- a union representative;

- experts from within the organization (for example, an on-staff occupational hygienist); and/or

- experts from outside the organization who have special expertise relating to some aspect of the incident (for example, the local fire chief, an outside occupational hygienist, or a trauma response specialist).

Where an investigative team numbers more than four members, it is a good idea to appoint one or two team leaders. These individuals can help bring order to the investigation and help keep it on track. They can also be the ones responsible for generating an investigation report and for completing any forms required by outside agencies (for example, a report to the MOL or a Form 7 for the WSIA, both discussed above). The team leaders should also be responsible for communicating the findings of the investigation to senior executives, to the injured worker, and to the worker's family, as well as to the other employees, where appropriate.

Investigation Objectives

Before beginning the investigation, investigators should have a clear understanding of the objectives of the investigation.

Research has shown that when it comes to preventing future harm, investigations that focus on prevention and learning from errors are much more useful than those that have as their primary focus the assignment of responsibility (or even blame) for the accident. The no-fault style of Ontario's worker compensation system has helped reduce motivation—on the part of employers and workers alike—to point the finger at each other in the wake of an accident. This is important, because many experts believe that searching for a single root cause (and an at-fault individual) creates a form of "tunnel vision" that stunts what should instead be a broader process of identifying system-wide causes (see Geller, 1996, p. 30).

Many accidents are caused by a complex convergence of circumstances, and striving to find a simple, single cause for every accident can hinder prevention efforts. The authors of *Safety Ethics* have the following to say regarding investigations:

> Deviations in individual and group work behavior ... are typically shaped by the unintended effects of changes in tools, processes, tasks, resources, interpersonal dynamics, or other factors ... with rare exceptions, ending an investigation by blaming and firing individuals for their involvement in a violation, near miss or adverse event will leave unsafe conditions in place to harm someone else. (Patankar et al., 2005, p. 75)

In reporting the findings of a post-incident investigation, investigators should be sure to describe *all* of the factors that contributed to an accident, so that the incident can be understood to the fullest extent possible. The report should conclude with recommendations for changes, where appropriate, that address each of these contributing factors.

The Importance of Follow-up Actions

Of course, simply reporting on an investigation is not enough. In order to support prevention of future accidents, the employer (and sometimes workers, too) must take appropriate and specific actions to remedy risks and problems. These actions might include:

- repairs to or replacement of equipment;

- the introduction of new PPE;

- changes to work processes, work organization, work flow, staffing arrangements, or other aspects of work;

- changes to the premises (for example, adding handrails, changing the direction in which a door swings, and so on);

- changes in the ways in which hazardous materials are handled;

- additional training for workers or supervisors; and sometimes

- disciplinary actions to enforce safety practices that have not been followed, or—far better—the introduction of new incentives to obey safety procedures.

When workers can see for themselves that employers are taking steps to remedy safety problems in the workplace, it reinforces the message that safety is taken seriously in the workplace and supports workers' own efforts to follow safety procedures.

KEY TERMS

grievance

essential service

REVIEW QUESTIONS

1. List at least three reasons why an appropriate and timely reaction by the employer to workplace health and safety incidents is important.

2. Why is it important, from a health and safety perspective, for employers to investigate environmental and/or human rights incidents in the workplace?

3. From the perspective of workplace safety culture, why would a MOL inspection conducted in response to a worker complaint be a "red flag" for the employer?

4. If a worker refuses work on the grounds that he believes it to be unsafe, and management—after investigating the concern—disagrees, what happens next?

5. Why should employers document and investigate near-miss incidents?

6. Under the OHSA, under what circumstances must an accident be reported to the MOL?

7. Under the WSIA, under what circumstances must an accident be reported to the WSIB?

8. Does the WSIB recognize (and compensate workers for) occupational illnesses that do not appear on schedules 3 or 4 under the WSIA?

9. Are employers required by law to provide trauma counselling services to workers in the wake of potentially upsetting events?

10. What should be the primary objective of a post-incident investigation?

AT ISSUE: IS THE INVESTIGATION OF WORK REFUSALS AN ESSENTIAL SERVICE?

At the time of writing, many commentators have expressed their concern that funding cutbacks have left the MOL without adequate human resources to fulfill its duties to the extent contemplated by the OHSA.

In 2005 the OLRB was asked to resolve an issue that had arisen in collective bargaining between OPSEU and the Ontario government (which determines staffing levels at the MOL).

OPSEU and the government were at odds over the minimum number of inspectors that should be required at MOL offices. The government contended that the appropriate minimum staffing level for inspectors was 23 (spread across 19 regional MOL offices), while OPSEU believed that the minimum should be 56.

OPSEU's position was grounded in its assertion that the investigation of work refusals is an **essential service** (the government is obligated to maintain enough staff to carry out essential services in a timely manner). The government disagreed.

After considering provisions in the OHSA that allow other workers to undertake work that one of their colleagues has refused, the OLRB found that, because the refused work could pose a legitimate danger, MOL investigation of work refusals had the potential to prevent serious injury or death, and hence is an essential service.

The OLRB ordered that the minimum staffing level for inspectors across the 19 MOL offices was 30—26 construction and industrial inspectors, plus 4 mining inspectors.

Source: *Ontario v. Ontario Public Service Employees Union*, 2005 CanLII 16183 (OLRB).

essential service
a service formally designated, by a government authority (a municipality, province, or the government of Canada) as an essential service

1. Do you agree with the OLRB's decision? Why or why not?

2. If the practice of a worker stepping in to do work that another worker has refused carries a potential for serious injury or death, why does the OHSA allow this practice?

3. Do you think that this kind of legal action (bringing an application before the OLRB) is an effective means of challenging government decisions, such as decisions about Ministry staffing levels? Why or why not?

CHAPTER 10

Facing an Investigation or Prosecution

CHAPTER OBJECTIVES

After completing this chapter, you should be able to:

- Explain the process for appealing an order of the MOL.

- Understand the difference between an MOL inspection and an MOL investigation.

- Describe the circumstances under which it is advisable to obtain legal representation.

- Explain how an accident, worker complaint, or other incident in a unionized workplace can lead to a grievance under the collective agreement.

- Understand how best to respond to a WSIB investigation based on allegations of non-compliance or fraud.

- Understand how a workplace accident can lead to a criminal prosecution or a coroner's inquest.

INTRODUCTION

In an ideal work world, the only accidents that would happen would be those that could never have been foreseen. When they did occur, employers would react immediately and appropriately, injured workers would receive prompt and fair compensation, repairs and preventive steps would be executed without delay, and—apart from receiving a report (if necessary)—the MOL would have no reason to become involved.

In the real work world, the IRS occasionally works imperfectly, and an external regulator (usually the MOL or the WSIB, but sometimes the police or a civil court) becomes involved.

The most benign kind of outside involvement an employer may encounter is a visit by an MOL inspector. The most serious kinds include MOL prosecutions, WSIB fraud prosecutions, and criminal prosecutions.

In between these extremes fall a range of events, all of which require action and strategic thinking on the part of the employer. In the previous chapter, you learned about the employer's responsibilities following an internally managed workplace incident. This chapter introduces external regulatory actions and appropriate employer responses.

While all workplace accidents are unfortunate—particularly when they result in injuries to workers—accidents that attract the attention of outside regulators have the potential to impact the company's reputation in the industry and in the community. For many companies, its reputation is its most valuable corporate asset. To avoid negative publicity, employers must make safety a priority, which means devoting adequate resources to safety compliance, even when budgets are tight:

> Tightening budgets exert pressure not only on staffing levels, but also on the resources available to support the occupational health and safety system. Professionals who feel they already have too much to accomplish in too little time may, if not adequately supported, feel overwhelmed by the burden of attending to health and safety compliance. Indeed, the array of legislative requirements for compliance not only with health and safety rules but also with environmental protection regulations is growing ever more complex. However, given the importance of the right to work in a safe and healthy environment, the importance of allocating resources to regulatory compliance cannot be overstated. (Rock & Campbell, 2002, para. 1.13)

MINISTRY OF LABOUR INSPECTIONS

Why They Happen

inspection
in the OHS context, a visit to the workplace by a MOL inspector to assess compliance with the OHSA requirements

In some cases (more frequently in high-risk workplaces), MOL **inspections** occur at random, as a function of the MOL's routine regulatory powers. In other cases, an inspection is triggered by a call to the MOL by a concerned party. This party may be either a worker or someone in a management position. An example of a situation in which an employer representative may place a call to the MOL is where a worker has refused allegedly unsafe work and has elected to continue to refuse to perform the work following the employer's inspection and remediation actions. This situation requires an investigation by the MOL, and the inspection is initiated by the employer.

In other cases, a call to the MOL may come from a worker. For example, where a worker has reported a problem to a supervisor and the supervisor has refused to fix the problem or communicate the worker's complaint to a higher-up, the worker may decide to contact the MOL for advice, or to complain. Complaints received in this manner are kept confidential by the MOL.

MOL inspectors are entitled to enter workplaces to conduct inspections without producing a warrant or providing the employer with notice that they are coming.

FACTS AND FIGURES

Are Some High-Risk Workplaces Being Overlooked by Regulators?

In January 2007 the CBC released a report that suggests that certain workplaces not traditionally viewed as particularly dangerous may have compensation claims rates as high as traditionally "dangerous" industries like forestry or mining. However, the CBC's report, which was compiled using information gathered through the access to government information process, reveals that regulatory bodies—like the MOL in Ontario—tend to focus their inspection program on construction, manufacturing, mining, and forestry. Workplaces such as offices and health care facilities are visited infrequently.

The CBC study points out that nurses, in particular, face a high rate of workplace injury. An analysis of Canada-wide statistics revealed that 32 percent of all nurses involved in direct patient care in hospitals and care facilities were assaulted on the job in 2005. Nurses also have a high rate of other kinds of injuries, and yet their workplaces are not inspected nearly as frequently as are, for example, construction sites. Data from British Columbia revealed that nurses' workplaces were up to 20 times less likely to be inspected than were forestry workplaces. The CBC study concluded that inspection practices by regulatory bodies are not in sync with workplace realities, but rather are influenced by outdated perceptions of which kinds of work are "dangerous."

The situation in Ontario may be somewhat better, however, than the national average. In 2004 the government of Ontario launched an injury-reduction initiative that included the hiring of 200 new OHS inspectors. One goal of the program was to reduce the lost-time injury rate by 20 percent by 2008; in April 2007 the Ontario Ministry of Labour reported that it was on track to reaching that goal, claiming to have prevented over 30,000 lost-time injuries in the first three years of the initiative.

What to Do

Workplace inspections are mandatory. An employer must permit entry to MOL inspectors, even if they arrive unannounced. Failure to do so can result in the employer being charged with an offence under the OHSA and fined.

Where an employer *is* aware that an inspection is going to take place, it can take several steps to prepare. As mentioned in the previous chapter, the employer should arrange for an employer representative to accompany the inspector. A representative of the workers is also entitled to accompany the inspector. This person will likely be a JHSC member or the Health and Safety Representative. If none of these people are available, it may simply be a knowledgeable worker (usually chosen by the union, if the workplace is unionized).

The employer should also be prepared to give the inspector access to a wide range of documents. These may include:

- blueprints/maps of the workplace;
- the workplace's health and safety policy, and individual policies created to govern specific work activities;
- documentation supporting tradesperson certification or licensing;
- first aid records;
- documentation used in support of the WHMIS (for example, MSDSs);

- inspection records for equipment;
- accident reports;
- training program materials;
- test results (for example, relating to air-quality testing or substance toxicology); and
- safety reports (for example, ground stability reports for a mine).

In some cases, inspectors may take objects, samples, or documents away from the workplace for further examination.

Employers are required to cooperate with inspectors by showing them around the workplace, asking questions, allowing photographs, and permitting access to workers for interviewing purposes. All employees who assist in an inspection are entitled to be paid at their normal rate of pay for any time spent helping the inspectors.

Inspectors' Orders

When an MOL inspector discovers a contravention of the OHSA or a regulation made under it during the course of an investigation, the inspector will make an order requiring the employer to achieve compliance. A copy of the order must be given to the Health and Safety Representative or to the JHSC, and a copy must be posted in an area of the workplace where workers will see it.

For minor contraventions that do not pose an immediate threat to safety, the employer will generally be given a time frame within which to comply. For problems that require considerable remedial efforts (for example, remodelling or remediation of an air-quality problem), the employer may be asked to submit a plan outlining the steps and timeline that it will take toward fixing the problem.

Where a health and safety contravention poses an immediate safety risk, the employer will typically be asked to comply immediately. The employer may also be ordered to stop all work in the affected area; to stop work that requires the use of an unsafe thing (for example, a toxic material, an unguarded machine, a process for which there is no safety policy, and so on); and/or to evacuate or cordon off the area.

In cases where the contravention poses an immediate risk, the employer will typically not be allowed to permit work to resume until:

- the MOL has been notified that compliance has been achieved; and
- a Health and Safety Representative or worker member of the JHSC has confirmed that, in his opinion, compliance has been achieved.

In the event that the employer defies a stop-work order, an MOL Director can request a court order from the Ontario Superior Court of Justice requiring the employer to stop work in the unsafe area.

In the case of any MOL order (not just those that relate to imminent risks), the employer is required to confirm compliance (within three days of achieving compliance) by sending a written notice of compliance to the MOL. This notice must be signed by the employer and must be accompanied by a statement by either the Health and Safety Representative or a worker member of the JHSC stating whether or not she agrees with the employer's claim of compliance.

Like the MOL order, the notice of compliance must be posted in the workplace in a location where the workers are likely to see it, and must remain posted for at least 14 days.

Appealing MOL Orders

MOL orders are appealable, which means they can be challenged. A party who wishes to appeal an order must do so within 30 days of the order being made. The parties eligible to appeal an order are:

- the employer;
- a constructor;
- a licensee;
- an owner (of work premises);
- a worker; or
- a trade union.

The appeal of an MOL inspector's order is heard by the OLRB. When one appellant (someone from the list above) launches an appeal, other people automatically become parties to the appeal, which means they have the right to make submissions to the OLRB regarding their views on the order. When an employer brings an appeal, the other automatic parties are:

- the workers;
- the union (if any); and
- the inspector who made the order.

If a worker or union brings the appeal, the employer is automatically a party.

To launch an appeal, the party wishing to challenge the order (from this point on, we will assume it is the employer) must fill out a form (Form A-65) provided for this purpose by the OLRB. The form (and an Information Bulletin that explains appeals of orders) is available on the OLRB's website. However, before filing the form with the OLRB, the employer must prepare an appeal package, copies of which must be given to all the other parties involved (see the list above).

The appeal package must contain:

- a copy of Form A-65 (the appeal application form);
- a copy of the Information Bulletin;
- a blank copy of Form A-66 (the form that the other parties will use to respond to the appeal); and
- a filled-out copy of Form C-44 (the form that gives formal notice of the appeal).

Once all of the appeal packages have been delivered, the employer can file its Form A-65 with the OLRB. The employer cannot file by email, registered mail, or by fax, but can file in person, by courier delivery, or by regular mail. This must be done within five days of delivering the appeal packages, or the OLRB will terminate the appeal.

The application form requires the employer to provide the following information:

- its address and contact information;

- the name, address, and contact information of the employer's "representative" (often, a lawyer);

- the names, addresses, and contact information for the responding parties;

- a copy of the inspector's order and/or inspection report;

- an account, in words, of the circumstances that led to the order (or the failure to make an order, if this is the reason for the appeal); and

- an account of the employer's reasons for appealing the order.

Once an appeal is filed, the OLRB has the power to suspend the MOL inspector's order pending the outcome of the appeal. When this is done, the employer need not take steps to comply until after either the appeal is settled or the OLRB makes a decision about whether or not the order will stand. Generally, if the inspector has alleged that the contravention poses an immediate threat to workers, the OLRB will not suspend any orders designed to protect workers (for example, a stop-work order).

In many cases, the OLRB will assign a Labour Relations Officer (LRO) to look into the facts of the order and the appeal. The LRO and the parties will be given a report date by which the OLRB wishes to receive a report about the case. Before that report date, the LRO will often work with the parties to try to settle the issue without the need for a formal hearing. This process can either be informal (for example, telephone calls from the LRO to the parties) or can involve a scheduled mediation session. At the session, the LRO (who is trained in mediation) will act as mediator. The LRO cannot make a decision about the conflict, and will not act as an adviser to any of the parties.

If the settlement/mediation process fails to resolve the matter, the OLRB will set a date for either a consultation or a hearing. Once this date is set, the parties who have received appeal packages have up to 21 days before the consultation or hearing date to respond to the appeal by filling out their Form A-66. The response form allows respondents to provide contact information for themselves and their representatives (if any), to describe the circumstances that led to the order from their own perspective, and to explain whether (and why) they support or oppose the employer (or other applicant's) appeal of the order.

The OLRB offers two different tiers of proceedings for dealing with appeals of MOL orders. The more informal of the two is the consultation. According to the OLRB, the precise format of a consultation varies depending on the circumstances, but in general, no witnesses are called. Instead, the board chair or vice-chair takes an active role, questioning the parties, expressing opinions, and finally, defining and ruling on the outstanding issues, either at the end of the proceeding or in a decision issued afterward. The decision rendered by the chair or vice-chair in a consultation is binding on the parties.

The possible outcomes of a consultation are:

1. The OLRB can "exercise its discretion not to inquire further into the appeal," in which case the MOL order will stand.

2. The OLRB can dismiss the appeal on its merits. Again, the order will stand, but in this case the OLRB can be seen as endorsing the inspector's decision to issue the order.

3. The OLRB can grant the appeal, in which case the appellant will not need to comply with the order.

4. The OLRB can schedule the matter for a full hearing (before the OLRB).

If the outcome is one of the first three on the list above, that is the end of the matter; if the OLRB chooses to schedule the matter for a hearing, a date will be set for the hearing.

A hearing is a formal proceeding, governed by the OLRB's rules of procedure. These include rules of disclosure, which means that parties must provide copies of all of the documents they want to use at the hearing (for example, letters, reports, and copies of relevant case law decisions) to the court and to the responding parties.

Most hearings involve the calling of witnesses, sometimes by both sides. Before the witness evidence begins, the parties will be invited to give opening statements; after all the evidence has been presented, they will be allowed to make closing statements. Following the hearing (this can sometimes be days or even weeks later), the OLRB will issue a final and binding order about whether the MOL order will stand or whether it will be set aside.

It is also possible for a party (most often, a worker or a union) to appeal an MOL inspector's decision *not* to make an order. This type of appeal is also brought before the OLRB. The opposing workplace party (usually, the employer) is automatically a party, and the appeal proceeds in the same way described above, with one difference: at the end of the process, the OLRB has the option to make an order requiring the employer to comply with the OHSA (even though the MOL inspector declined to make such an order).

Legal Representation

A party seeking to appeal a MOL order or the lack of an order, or a party responding to such an appeal, may choose to obtain legal advice to assist in the process; if the matter proceeds to the hearing stage, the party may choose to be represented by a lawyer before the OLRB.

Very large employers typically have lawyers either on staff or on retainer and available for consultation as issues arise. These employers will likely seek advice from counsel about whether an appeal is worth pursuing in the first place, and will choose to have a lawyer participate in the hearing if there is one.

However, because the OLRB's procedures (up to the hearing stage, at least) are designed to be quite accessible and informal, even an employer with counsel may choose not to involve its lawyer in the mediation and consultation stages of the process.

For this reason, parties—and responding parties in particular—who have no lawyer on retainer can proceed through to the end of the consultation on their own, and with reasonable confidence. This is especially true for a responding party whose views are closely in line with the views of the MOL inspector, because the inspector will have legal counsel available to support her arguments in favour of her order (or decision not to make an order).

If the consultation fails, most parties will want legal representation for the formal hearing before the OLRB, because this hearing requires strict adherence to the OLRB's procedures, and may involve tasks unfamiliar to people with no legal experience (for example, subpoenaing unwilling witnesses, analyzing case law, or cross-examining witnesses).

Parties seeking to minimize legal costs, however, should not overlook the benefits of having a lawyer assist in drafting the application or response forms. Taking time at the beginning of the process to properly identify the issues, the relevant facts, the reasons for appealing (or for opposing an appeal), and the desired remedy will make it more likely that the matter will be resolved at an early stage (in mediation with the LRO, for example), which can save money in the long run.

MINISTRY OF LABOUR INVESTIGATIONS AND PROSECUTIONS

Inspections and Investigations: What's the Difference?

MOL inspections can occur at any time, whether or not there appears to be a problem in the workplace. However, where a visit by a MOL inspector is prompted by a serious or fatal accident, an "unusual occurrence" (for example, a serious spill or fire), or a refusal to work, the inspection may be elevated to the status of an **investigation**.

> **investigation**
> regardless of the word used to describe the visit by the MOL, any "inspection" triggered by an OHS event (accident, spill, claim, etc.) is an investigation; inquiries by the MOL into potential fraud or other employer offences are also investigations

The key difference between an inspection and an investigation is that, while an inspection can lead to MOL orders (discussed above), an investigation can prompt a prosecution—where the MOL brings the employer before the court. A successful prosecution can result in a fine of up to $25,000 for an individual or $500,000 for a corporation.

When the MOL chooses to prosecute an employer, it is usually because the employer has committed a serious contravention of the OHSA that has resulted in a critical or fatal injury to a worker. One of the reasons for prosecuting employers is that a successful prosecution generally leads to a fine, which can deter the employer from future contraventions. Another reason is that a public denunciation of the employer's violation of safety laws can deter *other* employers from breaking the law. The results of prosecutions are widely available to the public, whether on the Internet, directly from the MOL, or in the general media.

From a "what happens" perspective, there was once little practical difference between an inspection and an investigation. In both cases, the MOL inspector was entitled to arrive without notice, to examine the workplace, to question workers, to view and copy documents, and to make orders—just like in the case of an ordinary inspection. However, while that is still technically possible under the wording of the statute, there is now some controversy about the validity of this approach where the visit is an investigation, and the MOL now chooses to obtain search warrants in some cases (see the Hard Lessons feature below for details). A search warrant may be drafted to allow the inspector to search and seize equipment, other physical evidence, and/or documents; however, it does *not* authorize the inspector to question workers or company representatives.

HARD LESSONS

When Is an Inspection Really an Investigation?

In 2001 the Ontario Court of Appeal released a decision that examined a statutory prosecution. *R v. Inco Ltd.* (2001) was not about an MOL prosecution, but rather about a Ministry of the Environment (MOE) prosecution. However, there are broad similarities in investigations under the MOE and MOL systems: for example, in the wake of an accident or incident, investigators from the MOE arrive at the workplace with significant investigative powers and begin collecting samples, taking photographs, and asking questions.

The investigation that led to the Court of Appeal case took place at Inco, a very large nickel mining company in Sudbury, Ontario. In the spring of 1993 a worker detected a leak of waste water from the complex into a neighbouring waterway. Tests of the water showed very high concentrations of contaminants, including nickel and iron. A worker reported the spill to the MOE, and investigators arrived at the complex and began their investigation. As part of the investigation, workers were asked to make statements, and were told that they were under a statutory obligation to answer questions. The evidence collected in this way was later used by the MOE when it prosecuted Inco for polluting a waterway and failing to immediately report a spill.

Inco appealed the charges, and in its appeal alleged that the MOE's actions in questioning employees amounted to an abuse of process. The court summarized Inco's concerns as follows (at paras. 17–18):

> [W]hen the IEB [Investigation and Enforcement Branch (of the MOE)] Officer engaged in this investigation, he already had reasonable grounds to believe that an offence had been committed and improperly used the inspection power under s. 15 of the OWRA [*Ontario Water Resources Act*] to build a case for prosecution. ... In engaging in such conduct, the IEB Officer infringed Inco's right under s. 8 of the *Charter of Rights and Freedoms* to remain secure against unreasonable search and seizure and its right under s. 11(d) to a fair trial.

The Court of Appeal agreed that *Inco* had a right to be concerned about this issue. Citing an important Quebec case, the court explained that there is a difference between an inspection and an investigation: if the regulatory officers have "reasonable and probable grounds" to believe that there has been a contravention of the legislation before arriving at the workplace in question, the visit is not an inspection but an investigation.

Because it was not clear in the *Inco* case whether the inspector had reasonable and probable grounds to believe that there had been a contravention before attending, or whether he simply had some information (falling short of reasonable and probable grounds), the case was sent back to the trial court level for a new trial. However, the court made it clear that, if the new trial revealed that the inspector had reasonable and

> probable grounds before he began questioning the employees, he should have obtained a warrant before beginning his questioning.
>
> Because the MOE and MOL regulatory schemes are fairly similar, the Court of Appeal's findings in this case would likely apply in the same way to investigations under the MOL. In other words, "inspectors" who attend a workplace in the wake of an accident and who have reasonable and probable grounds to believe that there has been a violation of the OHSA should obtain a warrant before collecting evidence.

The introduction of search warrants into the OHSA investigation process is only one of the reasons that these investigations have shifted, in recent years, from a cooperative tone to a more confrontational one. The other reason is that, since the changes to the *Criminal Code* introduced by Bill C-45, there have been instances in which a regulatory investigation (the MOL investigation) has led to a "parallel" criminal investigation with charges under the *Criminal Code*. (In fact, in most cases where there has been a serious workplace accident—and in all cases where there has been a death—the police will visit the workplace, often ahead of the MOL inspectors; read more about criminal investigations below.)

So far, this parallel prosecution scenario has been relatively rare. However, the idea that a regulatory inspection—in which employers have traditionally been expected to cooperate—can lead to the disclosure of information that might later become criminal evidence against the employer is problematic from the standpoint of self-incrimination. While the legal subtleties are beyond the scope of this book, employers should at least be aware that (especially in the case of a serious accident) cooperating with an MOL investigation has risks that are best assessed by the employer's lawyer, if he or she is available for consultation.

What to Do

When an MOL officer arrives in the wake of an accident and does *not* have a warrant, the employer has a decision to make. The employer can either allow the officer to conduct the investigation (just like in the case of an inspection, and including allowing workers to be questioned), or it can try to deny entry to the MOL officer on the ground that the officer requires a warrant (because the visit is an investigation). However, where it is uncertain whether the employer will be prosecuted or not, denying entry to the MOL officer might later be interpreted as an admission of fault by the employer. Determining what to do in the case of a no-warrant inspection is best left up to legal counsel, if the employer has a lawyer available at the time.

Where the MOL inspector does produce a search warrant, the employer is entitled to examine the warrant before allowing the inspector into the workplace. The warrant should specify what (areas, equipment, documents) the inspector will want to examine. The employer can advise workers that they are not required to answer any questions asked by the inspector. If the workers choose to answer voluntarily, however, their answers can be used against the employer in court.

Because an investigation (unlike an inspection) occurs in response to an event, the employer will know ahead of time that a MOL inspector is coming. To prepare for the inspection, the employer should:

- first, provide immediate assistance, first aid, and emergency transport to victims;

- evacuate and secure the accident scene;
- if there has been a fire, collision, spill, or other environmental incident that might have an impact on public safety, infrastructure, or natural resources, contact relevant public agencies (for example, the police, fire department, or the MOE);
- inform the JHSC or Health and Safety Representative, and the union, if any, and arrange for an appropriate person to accompany the inspector;
- call or obtain legal counsel, briefly describe the incident, and follow any instructions received;
- call and brief public relations staff, if any;
- arrange for trauma counselling, if appropriate;
- send home workers who have been evacuated or who appear to be suffering from anxiety related to the incident;
- request that workers who may be useful to the investigation remain on standby; and
- arrange for the availability of any documents that may be necessary to the investigation, but if possible, speak with legal counsel before handing them over. There may be situations in which an employer can hold back certain documents (notably, its own internal investigation report about the incident) on legal grounds.

Thinking Safe

Accident Response Plans

While not all accidents can be foreseen, the fact that there may be an accident at your workplace somewhere down the road *is* foreseeable.

In the wake of an accident, a certain amount of turmoil is to be expected. However, once the dust settles (and preferably, before the police and/or MOL arrive), certain key staff should be ready to take appropriate, rational action designed to minimize the fallout from the crisis.

The lawyers at Stringer Brisbin Humphrey, a Toronto law firm that specializes in employer OHS law, identify the key goals (after first aid and harm prevention) in the wake of an accident as follows:

- Opportunities to protect legally privileged documents are used;
- All appropriate legal requirements and investigators' demands are met, without incriminating individuals or the organization more than necessary;
- Details and evidence that could assist with a defence are preserved; and
- The rights of organizations and individuals, which reach heightened status during a criminal investigation, are known and protected. (Edwards, 2004)

In order to achieve these goals, Stringer Brisbin Humphrey suggests that employers have an accident response plan in place. This written plan, which should be developed and revised in the normal course of creating an OHS program, should describe the full gamut of post-accident responses, including not only accident investigation and statutory reporting procedures but also "confidential instructions for key frontline personnel to properly protect the interests of the corporation, its supervisors, and its officers and directors." All front-line personnel who will be responsible for implementing the plan should be thoroughly familiar with its contents and their roles.

For more about accident response plans, see the Stringer Brisbin Humphrey website at www.sbhlawyers.com.

When securing the accident scene, the employer is required not to disturb the scene *except* where doing so is necessary to:

- save a life or relieve suffering;
- maintain the operation of a public utility or transportation system; or
- prevent unnecessary damage to equipment or adjoining property.

When it comes to preventing unnecessary damage, employers should err on the side of non-interference with the scene in order to avoid future allegations of tampering with evidence. All non-essential personnel should be kept away from the accident site at all times.

Where it is likely that the investigation will lead to a prosecution, the employer may wish to take detailed notes describing the investigation, including whether the inspector produced a warrant, which documents were reviewed (not just which were copied and taken away), and who was questioned. In the event that the employer later forms the opinion that the inspector should have had a warrant, or that the warrant that she did have was flawed or improperly carried out, the employer will be able to support this belief using the notes.

The appropriateness of the employer's response in the aftermath of an accident and the willingness of the employer to cooperate with the investigation will often have a bearing on the fine imposed at the end of a successful prosecution.

AT ISSUE

Can OHS Defendants Withhold Documents on Privilege Grounds?

General MOL inspection powers include the power to request access to workplace documents relating to occupational health and safety.

One kind of document that relates to health and safety and that may be created in the wake of an accident is an internal accident investigation report. Safety-conscious employers will likely conduct an investigation as soon as possible after an accident, in an effort to identify causes and opportunities for preventive actions. The report generated by such an investigation may contain candid conclusions about mistakes on the part of the employer that contributed to the accident. Clearly, a document of this nature could provide highly prejudicial (harmful) evidence against the employer if it were used by the MOL in a prosecution.

Can the employer create an investigation report but keep it out of the hands of the MOL? Recent decisions suggest that the answer is "yes."

After a serious accident at Bruce Power's B generating station, lawyers for Bruce Power suggested that the employer prepare a report detailing its internal investigation of the accident. Bruce Power was instructed to label the report "solicitor–client privileged and litigation privileged." The report was delivered to the lawyers to help them in formulating a defence in case the company was ever prosecuted. The report was also disclosed to the union representing Bruce Power workers (and in fact, union representatives had assisted in the investigation). All individuals who received a copy of the report were instructed to keep it private and to destroy it after reading it.

Despite these instructions, a representative of the union gave a copy of the report to the MOL, and the MOL sought to use it in the course of a prosecution against Bruce Power. Bruce Power objected, and a Justice of the Peace agreed with the objection, entering a stay of proceedings (stopping the whole prosecution). When the MOL appealed this decision to the Ontario Superior Court of Justice, that court agreed with the Justice of the Peace's ruling that the report was protected by privilege; however, it overturned the stay, ruling instead that the trial could proceed but that the MOL would not be permitted to use the report. The MOL tried to appeal again, but the appeal was dismissed: the court confirmed that the document was subject to privilege.

Solicitor–client privilege protects the confidentiality of documents prepared by clients for their lawyers after litigation has begun. **Litigation privilege** is slightly different in that it can protect documents made *before* litigation where it can be shown that the documents were created in circumstances where litigation was anticipated.

The *Bruce Power* decisions make it clear that employers should try to assert solicitor–client and/or litigation privilege over internal investigation reports in circumstances where there may be a prosecution. For more details on how to successfully assert privilege, see the article by Kate McNeill and Ben Ratelband (McNeill & Ratelband, 2007).

solicitor–client privilege
a legal concept that permits a party in a legal action (including a party facing prosecution under a statute) to withold or refuse to disclose documents generated between that party and his or her legal counsel

litigation privilege
a legal concept that allows a party in a legal action to withold or refuse to disclose documents prepared "in contemplation of litigation"—that is, in preparation for the legal proceedings

Preparing for and Attending Court

Employers facing prosecution under the OHSA should hire qualified legal counsel. The high fines chargeable under the legislation ($25,000 for individuals and $500,000 for corporations), as well as the potential for jail time for convicted individuals and the potential for damage to the company's public image make appropriate legal advice and a well-planned and rigorous defence essential.

The need for legal counsel in defending an OHSA charge means that the employer's primary role, once counsel has been secured, will be to work effectively with the lawyer or lawyers. This will likely mean:

- discussing the circumstances relating to the incident openly with the lawyer;
- discussing the company's past health and safety record (including the company's history of compliance or non-compliance with MOL orders);
- providing the lawyer with access to documents relevant to the incident and to the employer's health and safety policy; and
- making witnesses available for interviews with the lawyer, and sometimes, for court appearances.

Lawyers preparing for the defence of an OHSA prosecution will grapple with a range of issues, for example:

- determining whether the charges can be challenged on procedural grounds (for example, whether they have been filed within the 12-month limitation period);

- determining how to plead (in some cases, it will be appropriate for the employer to plead guilty where the prosecution has a strong case. A guilty plea saves all justice-system participants time and money, and will generally be rewarded with a reduction in the sentence imposed);
- where it is appropriate to defend against the charges, determining which defences are available and what evidence will be necessary to support the chosen defence.

The defence most commonly asserted in response to an OHSA charge is **due diligence**. An employer who asserts due diligence makes the claim that it took all reasonable steps to avoid the incident that led to the charge. This is important, because in general, the MOL does not prosecute employers that have been compliant with orders, that have cooperated with inspectors, and that have an appropriate health and safety program in place.

There are two possible directions in which the employer can build a due diligence defence. The first is based on "belief in a mistaken set of facts."

An employer who relies on this defence alleges that it acted based on a reasonable belief in a mistaken set of facts and that, if the facts were true, the employer's actions would have been appropriate. For example, suppose that the employer received a new shipment of an innocuous product (for example, antibacterial hand soap) that had been used by workers on numerous occasions without PPE and without incident. However, on the occasion leading to the prosecution, suppose that the product received in the shipment was actually a volatile solvent, accidentally mislabelled by the supplier as hand soap and shipped in the containers normally used by the supplier for hand soap. When a worker opened the container and poured the solvent into a hand soap dispenser, the liquid splashed upward into his eyes and he suffered serious injuries.

The employer, in this case, would likely rely on the "belief in a mistaken set of facts" branch of the due diligence doctrine in its attempt to avoid a conviction.

The second—and more common—branch of due diligence is based on the concept of reasonable care. An employer who relies on this defence presents evidence to demonstrate that it took all reasonable precautions available to avoid the accident. The definition of "all reasonable precautions" will vary depending on the circumstances of the case, but may include:

- having an appropriate OHS program in place in the workplace;
- having appropriate training programs for employees, and having trained the workers who were involved in the accident;
- having a program of regular inspections for equipment in place, and having carried out inspections according to the program;
- compliance with the WHMIS for safe handling of dangerous materials;
- having appropriate PPE available, and having a track record of requiring workers to use it;
- having carried out the work activity that led to the accident according to a procedure that takes safety into account; and
- doing the work that led to the accident in an appropriate work environment (for example, in an area with sufficient ventilation).

due diligence
a possible defence against charges whereby a defendant asserts that it took all reasonable steps to avoid the incident that provoked the charge

As you might deduce from this list, establishing absolute due diligence is challenging, even for the most safety-conscious employer. The standard of proof in an OHSA prosecution is "on a balance of probabilities," which means that the prosecution need only prove that it is more likely than not (that is, 51 percent) that the employer was *not* duly diligent.

Defences based on anything other than due diligence rarely succeed. In particular, defences that seek to point the finger at third parties or to blame the accident on worker carelessness are ill advised. Protecting the safety of workers is the clear responsibility of the employer, and if the prosecution can establish that the accident could have been foreseen, a conviction is possible. If there were notable lapses in the employer's safety performance at the time of the accident, legal counsel may advise the employer to plead guilty.

Factors Influencing Fines

After a guilty plea or upon a conviction, the court must decide on an appropriate fine for each of the offences. As noted above, deterrence—both in relation to the convicted employer and to other employers—is an important factor that the court takes into account when determining a fine.

The court may also consider:

- the employer's ability to pay and the impact of the fine on the company (while a court will generally not impose a fine large enough to put a company out of business, it will also avoid levying a fine that would be seen by the public as a mere "slap on the wrist");

- the employer's past safety performance (including whether the employer has had previous convictions, has ignored orders, has acted in flagrant defiance of safety laws, or appears to have placed profitability ahead of safety); and

- the employer's response to the incident and to the prosecution (including whether the employer failed to take immediate steps to prevent future accidents, whether it failed to cooperate with investigators, whether it sought to blame others for the accident, and/or whether it mounted a defence that showed a lack of understanding of its responsibilities or otherwise demonstrated bad faith).

Minimizing Bad Press

When a serious accident has occurred in a workplace, media coverage is likely. Because the accident may lead to an OHS or criminal prosecution, there are potential legal implications for the company of disclosures made to the media.

However, refusing to answer *any* media inquiries can worsen the negative publicity flowing from the incident, because it can make the employer appear evasive, as though it is hiding something. It can also lead to the interpretation that the employer is unwilling to assume responsibility for the accident.

As part of the company's post-accident reaction plan, the employer should seek prompt legal advice about what information it can safely disclose to the media. Press releases or statements can be drafted and provided to the lawyers for approval. Having a brief but honest statement available for distribution to the press allows the employer to convey its sympathy for the victims as well as its commitment to finding and eliminating accident causes, while avoiding the risks of excessive disclosure.

It is useful to remember that, while the media may press for details of an accident, the public is probably less interested in specifics; instead, it is likely more interested in getting a general flavour of what has happened and an impression of the employer's reaction. Particularly in the wake of very serious accidents and fatalities, the media may be willing to wait for details when the employer cites the privacy interests of the injured parties or of their grieving families.

WSIB INVESTIGATIONS AND PROSECUTIONS

As you learned in chapter 5, the WSIA requires employers to report any material change in circumstance—for example, a change in business activities or the closing of a division of the business. Failure to report a change in circumstance is presumed, by the WSIB, to be intentional.

Where an employer has failed to report a change in circumstances, the WSIB can charge the employer with an offence under the legislation and commence an investigation and prosecution.

While the WSIB can take this action in response to any kind of failure to report, in practice the Board is more likely to charge and prosecute an employer who fails to report a change that would increase the amount of premiums it owes under the WSIA. A successful prosecution can result in the employer having to pay the difference in premiums paid and owing *plus* a fine, which can be up to $25,000 for an individual or up to $100,000 for a corporation.

Avoiding a WSIB prosecution is simple: the employer should ensure that its senior staff are trained to consider and act on WSIA compliance obligations every time there is a change in the way the company does business (that is, every time there is a physical move, a change in ownership, a change in business activities, or significant growth to the business).

WORKWELL AUDITS

In chapter 5, you learned that the WSIB conducts audits (described as "evaluations") of firms that seem to be incurring injuries at a higher rate than the average for firms in their rate group. These audits are conducted by Workwell, a WSIB agency. They serve the dual purpose of determining whether a firm should be assessed a higher premium than the norm for the rate group and providing guidance on ways in which companies can improve their safety ratings.

Workwell audits are not conducted at random. The WSIB has developed a "matrix" that it uses to assess firms and to set its own audit priorities. The matrix, which can be downloaded from the WSIB's website, is based on formulas that employers can apply to their own circumstances in order to determine whether or not they are likely to be targeted for a Workwell audit.

If an employer is worried about, or is faced with, a visit from Workwell, the employer should consider reviewing the *Workwell Audit* document available on the WSIB's website. The document provides an overview of the points that will be covered in the evaluation. Even if an audit is unlikely, applying the review described in the document as a sort of self-test can assist employers in identifying areas of concern. Taking steps to remedy those areas will make it very unlikely that the employer will fail an evaluation by Workwell, should one occur.

Before conducting an audit, the Workwell evaluator(s) will contact the employer in writing to give notice and to set a time for the visit. Workwell expects that at least two company representatives—the company owner or a senior officer, and a worker representative—will accompany the evaluator(s) on their tour of the workplace.

Once in the workplace, the evaluator(s) will go through the audit document, assigning points for areas in which the employer is in compliance. Employers who achieve a score of at least 75 percent of possible points are considered to have passed the audit.

Employers who achieve a score of less than 75 percent are given six months to achieve better compliance. At the conclusion of the six-month period, they are evaluated again. If they score below 75 percent at the second evaluation they are charged a penalty, which is calculated as a percentage surcharge on their basic premium; the percentage varies depending on the employer's score. The minimum surcharge is 10 percent, while the maximum (for the worst performers) is 75 percent.

GRIEVANCES IN UNIONIZED WORKPLACES

The OHSA gives trade unions a clear role in the IRS. Where there is a union in the workplace (the OHSA defines "trade union" in its Definition section; see chapter 3), the union is entitled to participate in the formation of the JHSC (and multi-workplace JHSCs), and to be involved in JHSC business.

For example, where a JHSC member complains that he has not been granted time to attend to JHSC duties or has not been paid for that time, the union can intervene on his behalf.

Unions are also entitled to receive information from the employer (for example, to be notified when there has been a death or critical injury) and to be contacted—as well as to assist in the investigation—whenever there is a work refusal. Union representatives may also accompany MOL inspectors when they visit the workplace, especially if JHSC worker members are not available to do so.

In general, the activities described above are routine union business and are performed without significant conflict with management. However, there are certain union powers that *do* give rise to a more adversarial posture for the union. One of these is the power to commence a grievance on behalf of a worker.

One of the most common grievances brought in the context of occupational health and safety is a grievance that alleges reprisals on the part of the employer. As you learned in previous chapters of this book, section 50 of the OHSA forbids employers from making reprisals against employees who exercise or seek to exercise their rights under the OHSA. In chapter 4, the Hard Lessons feature described what happened when a group of employees alleging reprisals and lost wages brought an application for redress before the OLRB. An alternative (and often, a simpler alternative) to that process—in workplaces where there is a union—is to bring a labour relations grievance before the OLRB with the help of the union.

When the issue of reprisals is addressed via the grievance process, an OLRB arbitration panel hears the grievance and arbitrates it by reference to the provisions of the collective agreement in place between the parties (the union and the employer). Because the collective agreement is designed to address the specific circumstances in the workplace, this process may be more efficient and satisfactory; for the worker, it is definitely a more accessible route to justice. For this reason, workers in a unionized workplace are somewhat more likely to challenge employer reprisals. Maintaining a harmonious—or at least a productive and fair—relationship with the union will make it easier for employers to navigate the grievance process.

CRIMINAL PROSECUTIONS

As noted above, in most instances where there has been a serious accident, the police are called to the workplace to investigate. A small percentage of these investigations lead to criminal charges.

The charge most likely to be laid by the police (unless the incident involved an assault or other interpersonal violence) is criminal negligence.

As you learned in chapter 6, criminal negligence differs from civil negligence in that, to be established, it requires a higher threshold of moral culpability. In other words, to convict an employer of criminal negligence, the trier of fact (judge or jury) will need to find—beyond a reasonable doubt—that the employer was reckless or incautious to a morally culpable degree.

Before the changes made to the *Criminal Code* by the passage of Bill C-45, a corporation could only be held criminally responsible for a workplace accident if the prosecution could attribute the harm that was caused to the negligent actions of the corporation's "directing minds." This was difficult to do, because a corporation's directors and senior executives generally do not provide immediate supervision to workers (there are usually several layers of management in between) and may not have direct knowledge of problems with equipment, flaws in work processes, and so on.

As you learned in the Hard Lessons feature in chapter 2, which discussed the Westray mine disaster, Bill C-45 overcame this obstacle to prosecution through two key changes to the law. The Bill also introduced *Criminal Code* section 217.1, which imposes an explicit duty of care on people who direct the work of others. Failure to discharge a duty imposed by law is, by definition, negligence.

Criminal charges have the potential for very serious consequences, and qualified legal representation is essential for an employer who is facing a criminal prosecution. Any time that the police are likely to attend at a workplace, the employer should contact counsel immediately. If possible, the lawyer should attend at the workplace so that he is available for on-the-spot consultation and so that he can observe any search or seizure. If counsel *cannot* attend or does not arrive in time, the employer should put a reliable staff person in charge of making detailed notes about the investigation. Observations that should be recorded should include:

- whether the police had a warrant (a copy should be obtained);
- which work areas were searched;
- what was photographed;
- which objects or surfaces were tested in other ways (for example, whether surfaces were tested for residues, whether samples of a spill were collected, whether temperature or other readings were taken);
- whether any equipment was operated by police, and how it performed;
- whether any equipment was moved or modified, and what state it was in before such testing (for example, whether a safety guard was in place when the police arrived, and whether the police moved it);
- what questions were asked of any personnel who escorted the police, and what answers were given;
- who (if anyone) was interviewed, by whom, and who was present, and whether any interviewees were cautioned (whether their rights were read to them);
- whether staff objected to any investigative actions (for example, whether any employee attempted to assert privilege over a document that the police wanted to see); and
- the observer should keep a detailed list of any items or documents that were seized by police.

A criminal prosecution involves a number of steps that are prescribed by the *Criminal Code*. These include:

- statement-taking and evidence-gathering;
- the formal laying of charges;
- bail hearings (if there have been arrests—though in workplace criminal negligence cases it is hard to imagine the court *not* granting bail, called "judicial interim release");
- disclosure (in a criminal trial, the prosecution must disclose all of its evidence to the defence; there is no reciprocal disclosure requirement);
- **plea bargaining** (in some cases), followed by the entering of pleas;
- pre-trial motions;
- jury selection (if there will be a jury);
- trial; and
- sentencing.

Legal counsel will guide the company through these steps. The lawyer(s) will need assistance or information from company officers at several stages in the prosecution and trial. It is a good idea for the employer to designate an individual to act

plea bargaining
an arrangement between a prosecutor and defendant whereby the defendant pleads guilty to a less serious charge with the expectation of leniency rather than be tried for a more serious one

as the primary contact for the lawyers. While this person may need to consult others within the organization to obtain the information needed, having one point of contact will avoid conflicting instructions being communicated to counsel.

One of the stages at which lawyer–defendant communications will be most important is at the plea bargaining stage. In some cases, where the prosecution's evidence (which will be shared with the defendant at the disclosure stage) warrants it, counsel will suggest a guilty plea. Pleading guilty avoids the need for a trial, which saves the court considerable time and reduces the legal expenses for both sides. In an appropriate case, a guilty plea will lead to a more lenient sentence. If the employer would have likely been found guilty anyway, it may also send a positive public relations message, because the employer may be perceived as being willing to take responsibility for its mistakes.

CORONERS' INQUESTS

Introduction

In some cases where there has been a workplace death, the Office of the Coroner examines the body of the deceased and prepares a report, called a coroner's report, which is filed at the Office of the Chief Coroner. In certain complicated cases, the coroner goes a step further and launches a fuller investigation that extends beyond the examination of the body and includes a hearing complete with witness testimony. This kind of investigation is called a coroner's inquest.

The coroner's office is a public office dedicated to the investigation of accidental or unusual deaths. There are several regional coroners' offices across the province of Ontario. The motto of the Office of the Chief Coroner of Ontario is "We speak for the dead to protect the living."

Coroners' reports and coroners' inquests have a fact-finding—rather than a fault-finding—purpose. Autopsies and inquests are conducted in an attempt to uncover the causes of unusual deaths, so that deaths under similar circumstances can be prevented in the future. Often, the filing of a coroner's report or the release of the findings of an inquest includes recommendations made by the coroner to individuals, agencies, or government departments. These recommendations suggest steps that may be taken in the future to protect the public.

The "verdict" in a coroner's inquest expresses the jury's findings regarding:

- who the deceased was; and
- how, where, when, and by what means she died.

Neither the jury nor the coroner makes any findings with respect to anyone's legal responsibility for the death.

What Triggers a Coroner's Investigation of a Workplace Death?

Section 10 of the *Coroners Act* requires that all accidental or unusual deaths be reported to the coroner. This includes all workplace deaths (while it is possible that certain deaths from natural causes may occur at a workplace, it is best to report all workplace deaths). The report can be made by anyone; typically, it will be made by the employer or by the MOL.

In some cases, the coroner will simply conduct an autopsy and issue a coroner's report; however, in certain circumstances the coroner is required by law to conduct a full inquest. Those circumstances are those in which a death results:

- from an accident on a construction project;
- from an accident in a mining plant, mine, pit, or quarry; and
- while the deceased was detained by a peace officer or was in a correctional institution or jail.

The first two circumstances clearly include certain workplace deaths. In the case of other workplace deaths—for example, a death in a factory—the coroner will decide whether or not an inquest is warranted. Finally, in some cases, the family of the deceased can request an inquest, and the coroner must consider the request.

What Is the Role of the Employer?

Where there is an inquest, the employer may be required to cooperate by allowing staff members to testify at the inquest. The employer may also be permitted to suggest witnesses who have not already been identified by the coroner, and may, in some cases, be permitted to ask questions. Where there is a union in the workplace, the union may also have the same responsibilities and privileges.

At the conclusion of the inquest, the employer may receive recommendations from the coroner regarding changes that should be made in the workplace. These recommendations do not have the force of law—that is, the employer cannot be forced to act on them. However, the combination of respect for the victim's family, a desire to protect surviving workers, and the need to minimize negative post-incident publicity generally motivates employers to make the recommended changes.

The government (in the case of a workplace accident, usually the MOL) has an additional responsibility: it must report back to the coroner about the steps it has taken or proposes to take in response to the recommendations. These steps can include the drafting of legislative amendments for submission to Parliament.

KEY TERMS

inspection

investigation

solicitor–client privilege

litigation privilege

due diligence

plea bargaining

REVIEW QUESTIONS

1. What triggers an inspection by the MOL?

2. As the result of an MOL inspection, an employer has been ordered to stop work on a key section of its assembly line until it can obtain a guard for a machine that lacks one. The employer has contacted the manufacturer and has learned that the guard is not in stock; it will take four weeks to obtain one.

    The employer knows of a way to bypass the manufacturing step that requires the guard's protection. The alternative is more labour intensive, but would allow the employer to continue production (stopping production, as the order requires, would lead to layoffs).

    What should the employer do?

3. Do all appeals from MOL orders require a formal hearing?

4. Why have some courts suggested that the MOL may require warrants to conduct an "inspection" that is triggered by a workplace accident?

5. List at least three steps an employer should take when it anticipates a post-accident visit by the MOL.

6. List at least four factors that the court will take into account when sentencing an employer convicted of an offence under the OHSA.

7. Why is it essential for an employer to report changes in the company's circumstances to the WSIB?

8. What happens if a workplace is audited by Workwell and fails the evaluation?

9. What does the prosecution need to prove, and with what degree of certainty, to secure a conviction for criminal negligence against an employer?

10. What is the purpose of a coroner's investigation into a workplace death?

CHAPTER 11

Managing WSIA Claims and the Transition Back to Work

CHAPTER OBJECTIVES

After completing this chapter, you should be able to:

- Explain an employer's responsibilities with respect to workers who are off work due to injury.
- Describe best practices with respect to the monitoring of injured workers and their claims for compensation.
- Explain the employer's responsibilities with respect to the reintegration of workers into the workplace after an injury.
- Understand the scope and limits of the duty to accommodate injured workers.
- Understand the legal issues with respect to layoffs and terminations of workers who have been away due to injury.

INTRODUCTION

The majority of workers who have been forced off work due to a workplace injury are eager to return to the workforce, and often to their original workplace. For most people, making a meaningful contribution to their community (in the form of work) and to their families (in the form of income) is essential to their self-esteem and quality of life. This is true of workers whether or not they have been partially disabled by their injuries.

Workers who are out of the workforce due to an on-the-job injury or illness and who are receiving benefits from the WSIA fund are still employees. Employers have a duty to manage these employees by ensuring that they are supported in their efforts to return to work.

Getting injured workers back to work is in everybody's interest. Workers who are off work cost the WSIA fund money, and their absence may affect their company's productivity. For the worker, being off work means less pay (WSIA loss of earnings (LOE)) benefits do not replace 100 percent of a worker's salary), and can sometimes contribute to feelings of depression stemming from the worker's inability to be productive and to do the things he could do before the injury.

This chapter will begin with a discussion of the legal and regulatory context for employers' return-to-work obligations, and will then move on to a discussion of strategic approaches and best practices.

LEGAL OBLIGATION TO PROMOTE RETURN TO WORK

Introduction

Sections 40 to 42 of the WSIA create a legal obligation, on the part of employers with 20 or more workers, to assist injured workers in making an "early and safe return to work" (ESRTW). The obligation only concerns those workers who have worked continuously for the employer for at least one year as of the date of the injury.

Employers are not the only parties with a role to play in the return-to-work process. Workers, workers' health care providers, the WSIB, and unions are all required to cooperate in the process. For workers, this usually means providing by the WSIB and the employer with reasonable access to certain health records. For unions, it means assisting the employer in making arrangements to accommodate returning workers. The WSIB provides guidance with regard to return-to-work and disability issues; assists, in some cases, with communications between the employer and health care providers; and hears complaints of non-compliance brought by workers.

This chapter will focus primarily on the employer's responsibilities, but will touch on the roles of other parties.

Assessing Return Date, Abilities, and Limitations

There are several components to the employer's obligation to rehire a worker. First, the employer must take steps to determine the likelihood of the worker returning to work, and the likely timeframe for her return.

Next, the employer must—with the assistance of the worker—develop an understanding of any temporary or permanent limitations that the worker may have as a result of her illness or injury. The reasons for this are threefold. First, the employer must determine what kind of work the worker will be able to perform on her return, including what she will and will not be able to do safely. Second, the employer must prepare to accommodate any special needs that the worker may have so that she can do her job. Finally, the employer must determine whether the worker's participation in the work will be safe for other workers around her, and for customers, clients, or other people with whom she may come into contact.

The WSIB has developed certain forms that can be used to assist in the process of determining what kinds of work a recovering worker can do. The employer, the worker, and the worker's health care giver should all be involved in filling out these forms.

PHYSICAL DEMANDS OF THE JOB

While the worker and his physician are best placed to provide information about the worker's injury or illness, recovery, and limitations (see below), the employer is best placed to determine the physical demands of both the worker's previous job and of other jobs in the organization that may now be suitable for the worker. The employer may wish to turn to this task while awaiting details from the worker.

The Physical Demands Information Form (PDIF) is actually a set of five forms (2828A, 2829A, 2830A, 2851A, and 2852A) designed by the WSIB to assist the employer in conducting a physical demands analysis of the worker's original job, and, if appropriate, of other jobs that are being considered for the returning worker. A full physical demands analysis is a fairly technical analysis that may require special expertise to perform; if a full analysis is required, the employer should assign it to either an outside expert or to an employee with training in this area. Completing the PDIF, however, is not a full analysis, but rather a simplified analysis that focuses on the risks of a job as they relate to a worker's particular abilities and disabilities. Generally, managers without special training will be able to complete this form.

The WSIB suggests that the ideal person to complete the PDIF is the worker's immediate supervisor, since he will likely have the greatest knowledge of the job tasks. In appropriate cases, the supervisor may seek input from the worker himself, even (in some cases) completing the form while watching the worker perform the job. Where this is not possible (for example, where the worker is still off work recovering from his injury and cannot do the job), the worker may need to complete his own copy of the form.

The PDIF form package includes the following:

- a guide to filling out the form (Form 2851A);

- a section 1 form (General Job Information, Form 2828A);

- a section 2 form (Job Duties and Physical Demands, Form 2830A, which requires the employer to break the job down into separate duties—for example, for a security guard, patrolling the workplace on foot would be one (or more) duties, and watching security monitors in the CCTV room would be a different duty. This form has separate sections for different body parts or body-part groupings: neck and shoulder; back; elbow, forearm, wrist, and hand; and hip, leg, knee, ankle, and foot;

- a section 3 form (General Physical Demands, Forms 2829A, 2829A2, 2829A3, and 2829A4, which has separate sections for different body parts or body part groupings: neck and shoulder; back; elbow, forearm, wrist, and hand; and hip, leg, knee, ankle, and foot; and

- a section 4 form (Additional Considerations for Injury Risk, Form 2852A).

Section 1 is compulsory for all employers. Depending on the nature of the employee's job, the employer can elect to fill out either section 2 (a more detailed analysis, involving dividing the job into duties) or section 3 (a general analysis). The guide suggests that while section 2 is always preferable, where a job involves a wide range of duties—more than six to eight duties—filling out section 3 instead would be appropriate.

Section 4 collects information about specific risks (for example, whole-body vibration or the need to work wearing thick gloves) that are present in certain workplaces only. Employers should review this form carefully (some of the risks listed are present in non-industrial workplaces) to determine whether any of the risks apply. If they do, the form must be completed.

Once the PDIF is complete, it should be submitted to the WSIB for the worker's file. It can later be useful for all the parties in determining the employer's and worker's compliance with ESRTW obligations.

WORKER'S ABILITIES AND LIMITATIONS

Once the demands of the job are understood by all, the next step is for the worker and his health care giver to complete an assessment of the worker's abilities and limitations. The document used to collect this information (though it should also be discussed more informally between the parties) is the WSIB's Functional Abilities Form (FAF). The WSIB makes this form, and a short guide to completing it, available on its website. The FAF is to be completed by three different parties: the worker, the employer, and the worker's health care giver.

The first portion, section A, of the form, which is completed by the worker and employer, asks for general contact information for the employer (including the listing of an individual who will serve as a contact for the purpose of managing the return to work), as well as for the worker's name, birth date, and accident date. There is also a small space provided for an answer to the question "type of job at time of accident." The short answer to this question should be supplemented, in most cases, by attaching either a full physical demands analysis or a WSIB PDIF (discussed above). The parties are also asked to indicate whether they have already discussed the worker's return to work. If they have not, they are expected to provide a date (in writing, on the form) on which they will have this discussion. This requirement is intended to encourage the parties to begin the process of planning for the worker's return as early as possible.

The next section, section B, serves the important function of indicating the worker's consent to the health care provider's release of the health information requested by the form. As you know, personal health information is closely protected by privacy laws, and a health care provider is not entitled to provide the worker's health information to the employer without the worker's consent. In signing section B, the worker provides his consent for the disclosure of a limited amount of health information: only that information necessary to help the employer understand the worker's post-accident capabilities and limitations as they relate to his work.

Section C requests billing information from the health care provider (so that the WSIB can pay certain health care bills on behalf of the worker, such as the fee for completing the form). Because this information is confidential, the health care provider should send a completed section C directly to the WSIB, leaving the section blank on any copies of the FAF provided to the worker or employer.

Sections D, E, and F are completed by the health care provider, and contain information about the worker's abilities. Section D provides the health care provider's overall assessment of whether the worker is:

- able to return to work with no limitations;

- able to return to work with restrictions on activities; or
- unable to return to work due to the need to recover further.

Section E elaborates on any restrictions that the worker may have. For example, part 1 of section E requires the health care provider to report whether the worker is capable of lifting objects from floor to waist:

- with full abilities;
- with a limit of 5 kilograms;
- with a limit of 5 to 10 kilograms; or
- "other."

Where the physician checks a box marked "other" or equivalent, she is expected to give further details. There is space on the form for additional comments, or she can attach a separate sheet.

At the end of the section that describes the worker's limitations, the health care provider is required to estimate how long restrictions on the worker's abilities are likely to be required. Answer choices include 1–2 days, 3–7 days, 8–14 days, or 14 or more days. The health care provider is also asked to indicate whether the worker can return full time, or on modified or graduated hours. The health care provider is asked to provide the date of the worker's next follow-up appointment so that the employer can have a sense of when the worker is likely to be assessed again.

Identifying Safe and Suitable Work

Once the employer has adequate information about the worker's condition and needs, it must determine what safe and suitable work is available for the worker to perform. Even if the worker is returning with no limitations on her abilities, the employer may have filled the worker's previous job with a new candidate (especially if the worker was away for an extended period of time). This does not excuse the employer from its obligations. If the worker can no longer do his previous job or if that job is no longer available, the employer must find suitable work for the worker, and that work must be paid at a rate "comparable" to the worker's former wage.

Finding a new job for a worker within the organization can pose special challenges for certain kinds of employers. Very small employers may simply have no positions available; this is the reason that the duty to support ESRTW does not apply in workplaces with fewer than 20 workers. Employers with staffs numbering just over 20 may need to reorganize the work of other employees to make room for the returning worker—a task which raises issues of accommodation (discussed below).

Employers of unionized workforces may face challenges related to the application of the collective agreement in force in the workplace. These employers will need to work closely with the union to identify suitable jobs for returning employees. The WSIA discusses the relationship between the employer's obligation to promote an ESRTW and the collective agreement at section 15:

> If this section conflicts with a collective agreement that is binding upon the employer and if the employer's obligations under this section afford the worker greater re-employment terms than does the collective agreement, this section prevails over the collective agreement. However, this subsection does not operate to displace the seniority provisions of the collective agreement.

In practice, this provision can prove to be quite ambiguous, because the most common scenario in which WSIA obligations conflict with the collective agreement is one in which the only suitable position that exists in the organization for a returning worker is one for which the worker has insufficient seniority. The employer will have to work closely with the union to find a way to reintegrate the worker into the workplace while complying as closely as possible with the collective agreement. Where there seems to be no way to do so, it may be useful to consult the WSIB for guidance.

Accommodating the Worker

Where a worker returning to work has a disability, the employer has a duty to accommodate the worker's special needs so that the worker can work safely and comfortably. Accommodations made for a worker can take many forms. For example, the worker

- may be given permission to work in a different way than usual (for example, he may be permitted to sit to do a job that is normally done standing);
- may be given extra breaks to allow her to rest or stretch;
- may require modifications to the work area to accommodate assistive devices (such as a wheelchair);
- may require special office equipment (for example, a kneeling chair, a standing desk, a special backrest or footrest, or a wrist support device);
- may require adjustments to conditions in the workplace (for example, a worker recovering from a brain injury may need to work in a quiet area);
- may need to avoid certain tasks that are normally part of the job (for example, he may need to be exempt from a requirement to restock products on a high shelf because he cannot lift objects overhead—these tasks may need to be assigned to co-workers);
- may need to work reduced hours (for example, because of lingering fatigue); or
- may need to work in a different area of the workplace (for example, in an area with excellent ventilation, because of respiratory problems).

There are many other possible accommodations.

In practice, it is generally up to the returning worker, with the help of his physician, to suggest appropriate accommodation measures. These suggestions should be communicated to the employer well in advance of the worker's proposed return date, to give the employer time to consider and implement them. The Canadian Human Rights Commission (CHRC) notes, in its 2007 publication *A Guide for Managing the Return to Work*:

> Supervisors are entitled to ask for additional time to assess an accommodation request if the employee doesn't provide enough notice prior to their expected return to work. For their part, employees have a responsibility to give their supervisors enough time to prepare for their return to work, particularly if they are requesting accommodation for specific needs.

What will constitute sufficient preparation time will depend on the kind of accommodation requested and the nature of the workplace. If the employee requires ramps to be built, for example, or requires a job reassignment that affects co-workers' jobs, several weeks' notice will be needed. If the employee's request is simpler—for example, if she requires additional rest breaks—less notice is required.

Regardless of the nature of the request, the employer should be given the time and the opportunity to consider the request and to investigate alternatives. The reason for this is that the employer is required to accommodate the employee, *not* to provide the *precise* accommodation requested by the employee. The employer may have legitimate concerns about the impact of the worker's suggested accommodation on several things, including:

- the worker's productivity;
- the employer's budget;
- the comfort or safety of other workers; or
- matters of seniority, whether or not there is a collective agreement in place in the workplace.

Cases decided under the WSIA have established that the employer's duty to accommodate disabled workers must be balanced with the employer's duty to protect its other workers, and with the employer's own legitimate interest in maintaining workplace productivity. Employers are expected to endure some cost, inconvenience, and loss of productivity while discharging their duty to accommodate. However, they are *not* required to place other workers at risk or accept workers' precise suggestions for accommodation where other accommodation options would be adequate and would have less impact on productivity.

Section 41(6) of the WSIA makes it clear that there is an upper limit on the sacrifices that an employer can be expected to make in the name of accommodation:

> The employer shall accommodate the work or the workplace for the worker to the extent that the accommodation does not cause the employer undue hardship.

There is no absolute objective guideline for what constitutes "undue hardship." The reason for this is that the circumstances of every case—and every employer—are different. Large, financially stable employers will suffer far less economic hardship from the purchase of special office equipment than will smaller offices. Larger employers may also be better able to tolerate a reduction in productivity (for example, due to an employee having to take frequent breaks) than would smaller workplaces where there may be fewer workers available to cover for breaks and absences. In addition, larger employers may also be more likely to be unionized, and the collective agreement in place may make it more difficult to reorganize the delegation of work (although it cannot be an absolute barrier; see below). Finally, the nature of the worker's injury (that is, whether it—and the accommodation measures—will be temporary) will also have a bearing on how much hardship the employer will be required to bear.

If the employer and the worker cannot agree on whether or not a requested accommodation amounts to undue hardship, the worker may bring the issue before the WSIB for resolution. If the WSIB finds that the accommodation in question does *not* constitute undue hardship and that the employer has not complied with the legislation, the employer may be levied a penalty. Because a failure on the part of the employer to comply generally means that the worker will remain out of work, the penalty for non-compliance is determined by reference to the worker's net average earnings (NAE) for the year preceding the injury. The maximum penalty chargeable is the equivalent of one year's wages at the worker's NAE. The WSIB will pay the penalty funds to the worker over time as though they were LOE benefits (see chapter 5 for a discussion of these benefits).

Thinking Safe

The WSIA Second Injury and Enhancement Fund

One of the concerns that employers may have when considering hiring a person with a disability related to a previous injury is that the previous injury may predispose the worker to a more serious disability, should she suffer a new injury at work. For example, if a person who lacks the full use of her left hand applies for a job and is hired, the employer may worry that if the employee should injure her *right* hand on the job, she will be off work completely, as she cannot use her hands. As a result, the employer will be saddled with a significant WSIA claim.

To offset the likelihood that these legitimate concerns will pose a barrier to the hiring of people with disabilities, the WSIA has created the Second Injury Enhancement Fund (SIEF). SIEF, which is available only to Schedule 1 employers, absorbs all or part of a second injury claim where that claim is enhanced or prolonged as a result of a previous injury.

SIEF is available even where the first injury occurred outside of the workplace and was not the subject of a WSIA claim (for example, where the first injury related to an automobile accident).

There are deadlines for applying for SIEF that can affect how the new injury will be treated under the NEER, MAP, and CAD-7 incentive programs (for more about these programs, see chapter 5). For more information on the SIEF program, employers can contact the MOL. There is also a good summary of the program on the website of the Office of the Employer Adviser at www.employeradviser.ca.

When Does the Duty to Support ESRTW Expire?

The obligation to rehire a worker following a workplace injury is not unlimited. The obligation expires at the earliest of three possible events:

1. the two-years-post-accident mark is reached (without the worker having returned to work);

2. one year has passed since the date at which the worker was deemed by a physician to be medically able to perform the duties of his pre-injury job (without the worker having returned to work); or

3. the worker reaches age 65.

This time limitation prevents the employer from being permanently "on the hook" to re-employ a worker who seems unlikely to be able (or willing) to return to work, and allows the employer to find a permanent replacement for that worker.

It is not possible for an employer to get around the obligations created by these provisions by rehiring a recovering worker, only to fire her soon after. Any termination of a returning worker that occurs within six months of the worker's return to work will be deemed, by the WSIB, to be a failure on the part of the employer to meet its obligations under the legislation.

FACTS AND FIGURES

Rehired, Then Fired?

Where a worker who has returned to work under the ESRTW provisions is fired within the first six months after his return, the worker is entitled to dispute the termination. To do so, the worker must file an objection with the WSIB within three months of the termination.

Once it receives the objection, the WSIB will consider the circumstances of the termination. Because any firing within six months of an ESRTW is considered to be a failure on the employer's part to meet ESRTW obligations, in order to have the termination upheld the employer will need to *rebut the presumption* that the firing was related to the worker's injury. If the employer cannot provide enough evidence to support a different reason for the termination, the WSIB will decide that the employer is not in compliance with the legislation.

If the WSIB decides in favour of the worker, the WSIB can charge the employer a monetary penalty. The maximum penalty is an amount equal to one year's worth of earnings at the worker's net average earnings level for the year prior to the accident. The penalty money collected by the WSIB in this way is paid out to the worker.

The Worker's Obligations

As is clear from the discussion of the FAF and accommodation preparations, above, the worker has a role to play in the back-to-work process. The WSIA states that workers must

- contact their employer after the injury, and maintain contact during the "period of impairment" (that is, while the worker is off work);

- cooperate with the employer to help "identify suitable employment that is available and consistent with the worker's functional abilities and that, when possible, restores his or her pre-injury earnings"; and

- give the WSIB any requested information that can assist in the process.

One of the key requirements for a worker's successful return to work is that the employer have sufficient information about the nature of the worker's injury. This will not be possible unless the worker is willing to consent to a limited disclosure of personal health information for the purpose of filling out the FAF. Also essential will be a certain amount of willingness, on the worker's part, to consider the accommodation options that the employer is prepared to offer, and to offer alternatives if none seems suitable.

Where a worker appears unwilling to cooperate in the process, the employer can contact the WSIB to express its concerns. If the WSIB agrees that the worker is not cooperating, it may penalize the worker by reducing or suspending her LOE or other benefits.

STRATEGIES FOR MANAGING THE RETURN TO WORK

Introduction

Now that the legal framework for the return to work has been explained, it is appropriate to consider practical issues. As noted by the CHRC in its 2007 publication *A Guide for Managing the Return to Work*:

> Return-to-work cases present challenges because they involve the careful balancing of an employer's right to manage a productive workplace with a worker's fundamental right to equality, dignity and privacy.

Employers who respond appropriately to workplace injuries and who take a good-faith approach to their duty to support the worker's return are much less likely to struggle with barriers to accessing essential worker information, inflexible accommodation requests, or other difficulties that can plague the return-to-work process. Maintaining good communication with the worker throughout the process is perhaps the single most effective strategy employers can use.

Post-Accident Communication

Employers and injured workers have an express legal duty to communicate with each other after an accident. Legal duty aside, however, employers who have a good relationship with their workers will naturally wish to inquire about the worker's health. Any such inquiry, however, will need to be made with tact, sincerity, and respect for the worker's privacy interests.

The person who is likely best placed to make the first contact with the worker is the worker's immediate supervisor. Before making a call to the worker, however, the supervisor should prepare by doing a number of things.

First, the supervisor should determine what kind of information each party is likely to need from the other. Following an accident, the parties will have quite divergent interests. The worker will be eager for information about how to make a WSIB claim, how to get reimbursement for medical costs, and how to access rehabilitation and healing programs and services. The supervisor should consult with people in the office (for example, staff in the HR or occupational hygiene departments) to gather as much information as possible, so that she can be a resource for the worker. At the very least, the supervisor should have contact numbers available for the WSIB and for any local injured workers' support organizations. Examples include the Workers' Health and Safety Centre (www.whsc.on.ca) or a local occupational health clinic (for locations, see the Occupational Health Clinics for Ontario Workers Inc.'s website at www.ohcow.on.ca).

The employer, on the other hand, will be eager to fulfill any reporting duties it has under the OHSA or WSIA, and to begin planning for the worker's return to work. The supervisor should consult with others in the organization to determine the precise information that will be needed from the worker. Making return-to-work plans will require two key pieces of information:

1. the worker's anticipated return date; and
2. the worker's capabilities and limitations.

Where the injury was a serious one, it will likely be both fruitless and tactless to request this information in the immediate aftermath of an accident. This does not mean that the worker should not be contacted; rather, it means that the supervisor should understand what is required so that she can plan for future communications with the worker.

Another step that the supervisor should take before making her first contact is to speak to the individual or individuals in the chain of command who will have the authority to make return-to-work offers and accommodations for the worker. These individuals may have instructions for the supervisor with respect to information that will need to be collected. They may also provide a message of support for the worker, which the supervisor can offer on their behalf. Alternatively, in organizations where the worker had regular contact with senior management, these individuals may choose to call the worker directly to express their support.

In summary, where the accident was a serious one and where a long recovery period is anticipated for the worker, the first contact with the worker will generally cover the following:

- an expression of personal support and concern for the worker's health and well-being, from the worker's supervisor, balanced with respect for the worker's privacy;
- a message of support on behalf of more senior executives in the workplace;
- the offer to provide any contact information, forms, and so on that may assist the worker in making his WSIA claim; and
- an assurance that the supervisor will contact the worker in the future to begin discussing return-to-work considerations.

Ideally, the mention of future conversations about return-to-work issues should include a proposed date for that call, for example: "May I call back again in a week (or "at the beginning of April," or whenever seems appropriate under the circumstances) to talk about how we can plan for your return?"

Where the accident was less serious—for example, where the worker has sustained a cut, sprain, strain, burn, or minor fracture—it will often be appropriate to begin discussing the worker's return to work in the first post-accident conversation. In this case, the supervisor may simply say something along the lines of: "To help us plan for your return to work, there's a form (referring to the FAF) that you can give to your doctor. May I mail you the form today?"

Regardless of the circumstances, return-to-work discussions must be handled with the utmost tact. The worker may be feeling vulnerable and afraid in the wake of the accident, and he may interpret the employer's inquiries as pressure to return to work before he is ready. To avoid this, the employer should ensure that the supervisor chosen to contact the worker has been well trained, and has good people skills and sensitivity. If she does not, another manager—perhaps one level above—or a person from the HR department may be better suited to communicate with the worker.

While pressuring the worker to return before he is ready is inappropriate, it is equally dangerous to make him feel isolated or ignored by failing to contact him early or frequently enough. The New Zealand *Aftermath* study discussed earlier, found a positive correlation between employers' post-accident involvement with workers and workers' successful rehabilitation:

> Support and its presence or absence was of considerable importance in the study. Participants often felt isolated from support structures. These included sources of information, support groups, and often, infrastructural support. This affected their ability to cope and their recovery. Related to this was the attitudes they encountered in their workplace. Attitudes and support received from the workplace affected the psychological and physical recovery process of individuals, as well as their ability to understand and handle their condition. ... In certain cases it impacted on the participant's family's ability to deal with the aftermath. (Adams et al., 2002, p. 11)

Several pages later, the authors of the study conclude:

> The approach taken to the injury event by the supervisor or employer impacted on the rehabilitation outcomes for the worker in two cases (Lisa and Grant). The most successful outcomes for both the worker and workplace were when the employer took an active role with the worker and appropriate responsibility for what occurred. This is the most evident in the case of Lisa, whose manager supported her and gave evidence at an ACC hearing. (Adams et al., 2002, p. 15)

In this last paragraph, "AAC" refers to the local agency that is equivalent to Ontario's WSIB.

Establishing supportive lines of communication between the employer and the worker as soon as possible after the accident will help eliminate surprises and conflicts when the time comes to negotiate specific accommodation solutions.

Planning for Accommodation

As the worker's planned return date approaches, the employer and worker must begin the process of negotiating the specifics of the accommodation that will be offered to the worker. As was explained above, there are a wide range of actions that an employer can take in the name of accommodation.

Where communications between the employer and worker are positive and productive, it is often useful for the worker to begin the process of identifying accommodations. The employer can invite the worker to make some suggestions. In order to reinforce the dual message that arriving at acceptable accommodations is not a one-sided process and that the employer is not obligated to make the precise arrangements requested by the worker, the employer should be careful about how this invitation is extended. The employer may say, for example:

> Could you please give some thought to how you can work comfortably after your wrist injury? Once you have come up with a few possible solutions, we can discuss them together to determine feasibility.

This approach has the advantage of communicating:

- the employer's willingness to consider the worker's suggestions;
- the employer's willingness to adopt a consultative/collaborative approach;
- the employer's expectation that the worker will be flexible (by providing alternative options); and
- the employer's prerogative to reject certain solutions.

Once the worker has made some accommodation suggestions, it is almost always best for the employer to request time to investigate those suggestions before rejecting or accepting any of them. The outright rejection of a suggestion can inject an unnecessary measure of conflict into the process, while a too-hasty acceptance can cause problems should it turn out that the suggestion is not feasible (for example, if it is more expensive than originally expected, if it violates the collective agreement, and so on).

The employer should review the worker's suggestions in good faith, with input—if appropriate—from experts on issues such as ergonomics or occupational hygiene. These experts may be able to propose alternatives that are easier to implement but that will be equally effective in promoting the worker's comfort. If the worker has made a request that has the potential to affect the rights of other workers (for example, a request to work on a different shift), the employer may need to involve union representatives in the decision-making process; however, if the accommodation requested will inconvenience other workers only slightly, these objections do not amount to undue hardship.

Informal discussions in which the worker and employer present alternatives and listen to each other's feedback may take place over the course of several conversations. However, the employer should eventually be the party to make a "final offer" with respect to accommodation. While doing so is not absolutely required by the legislation, it is a good idea to put the details of this offer into writing. Should the worker ever allege that the employer failed to offer accommodation or that the offer was unreasonable, it will be useful to have written evidence of the details of the offer.

Timing and Preparations

The written accommodation/return-to-work-terms offer should not only cover all of the details of the proposed accommodation (which may include, for example, a plan for graduated hours with target dates upon which hours will increase), but should also stipulate a return date for the worker.

The worker (with the advice of his health care provider) is generally entitled to choose his return-to-work date. The employer should do its best to learn this date from the worker so that an offer of accommodation can be made and accepted with enough time remaining for the employer to prepare the necessary accommodations. The preparations can include tasks such as:

- construction (for example, adding ramps, widening doorways);
- ordering and receiving supplies (such as anti-fatigue standing pads, special chairs, extenders for pedals and levers, and so on); and
- giving notice to other workers of changes to their jobs, and providing any training necessary to support those changes.

All of the aspects of the planned accommodation must be in place on the date that the worker returns to work.

Preparing Co-Workers

In some cases, the employer may anticipate some necessary adjustments on the part of the returning worker's co-workers. For example, they may need to take on additional tasks that were previously handled by the injured worker but that he can no longer perform, or learn a new procedure for completing a group task that involves working within the returning worker's new limitations. Sometimes, they may simply need to accept that the returning worker has been accorded what appear to be advantages (for example, extra rest breaks) that are not available to all workers.

People can be quite averse to change. Having to change the way one works to accommodate a colleague may seem unfair to some of the worker's co-workers. The employer can attempt to relieve some of this sense of unfairness by preparing the workers for the injured worker's return. A group meeting may be the best way to do this (in some cases, additional training will also be needed). At the meeting, the employer may wish to explain the following:

- the nature of the returning worker's injury, in basic terms, without violating the worker's privacy (to avoid privacy pitfalls, the employer should describe only those aspects of the worker's injury that will be visible to his co-workers;
- the employer's legal duty to support the worker in returning to work and to accommodate the worker;
- the steps that have been or will be taken to accommodate the worker;
- the expected duration of the accommodation measures, if they are temporary; and
- the employer's expectation that the returning worker's co-workers will cooperate with the return-to-work plan.

By explaining that there is a legal duty for the employer to accommodate the returning worker and by describing the accommodation plan in positive terms, as a plan designed to support the worker, the employer may send the message that it would be equally committed to accommodating any of the other workers should they ever find themselves in similar circumstances. This may help prevent complaints about unfair treatment. The employer should make it clear that it will not tolerate harassment or discrimination directed at the returning worker (or, indeed, at anybody).

Finally, once the returning worker has settled into the job, the employer should follow up with both him and his co-workers to ensure, in the returning worker's case, that he is comfortable and productive, and, in the case of the co-workers, that they are not overburdened by any additional job responsibilities.

COMPLAINTS AND PROBLEMS IN THE RETURN-TO-WORK PROCESS

The responsibility to effect an ESRTW is intended to be shared between three parties: the employer, the worker, and—to a lesser extent—the worker's health care provider. Only where the system has seriously broken down will the WSIB take an active role in the return-to-work process.

Some problems that may arise do not affect the employer directly. For example:

- the health care provider may make a determination that the worker is fit to return to work, but the worker may not agree; or

- the WSIB may make a determination that the worker is fit to work and no longer requires benefits.

In these cases, the worker may return to work but may fail to make a good-faith effort to succeed. The employer should be alert to what is going on in these circumstances, and should document any matters of concern (for example, events that cause the employer to suspect that the worker is intentionally not working safely or productively). Because any termination that occurs in the six months following a return to work can be challenged by the worker, the employer will want to be prepared to argue that the worker was terminated for good cause.

Problems that affect the employer more directly may include cases where the worker

- refuses to disclose needed information about his health, capabilities, and limitations, making it difficult for the employer to plan for his return to work;
- refuses to disclose a proposed return-to-work date, or even whether or not he plans to return;
- rejects the employer's offers with respect to the positions available to him on his return or with respect to accommodation options;
- insists on a particular accommodation plan that the employer considers to be disruptive, too expensive, or otherwise impractical; or
- returns but is unproductive or disruptive.

To the extent that it is practical, the employer should be patient with the worker. Sometimes, readjusting to work poses emotional challenges in addition to physical ones. Returning to work with a disability—even a temporary one—can be intimidating for the worker. The worker may not express this intimidation openly (or even recognize it, in some cases), but may reflect it in a less-than-cooperative attitude toward the return-to-work process. A certain amount of tolerance on the employer's part may encourage the worker to come around.

Where the worker is completely resistant to ESRTW goals, however, it is appropriate for the employer to contact the WSIB to explain the obstacles it is facing. In a case like this, the WSIB may be able to intervene by explaining the worker's responsibilities to him. At the very least, the WSIB will be alerted to the situation, and may begin making plans to assist the worker outside of the context of his original employment.

Where a worker believes that the employer is being uncooperative in the return-to-work process, the worker can raise his concerns with the WSIB. If the WSIB cannot prompt the employer to cooperate, it may fine the employer, as discussed earlier. If the worker is unable to secure a position at his original workplace, he may need to seek work elsewhere.

Sometimes, workers who are eager to return to their original workplace but whose employer is not cooperating may launch a complaint with the CHRC. This phenomenon is common enough that the CHRC has published a guide to best practices in promoting a return to work (the guide was mentioned earlier in this chapter). Where the CHRC makes a determination that the employer has discriminated against the worker on the basis of disability, the CHRC can make an order against the employer. A CHRC order can require, for example, that the worker be accommodated, reinstated, or compensated.

> # Thinking Safe
> ## The Fair Practices Commission
>
> The Fair Practices Commission plays an ombudsman role for workers and employers in the WSIB system. On its website at www.fairpractices.on.ca, it describes itself as follows:
>
>> The Fair Practices Commission is the organizational ombudsman for the Workplace Safety and Insurance Board. The Commission provides an independent, neutral and confidential service for injured workers, employers and service providers with complaints about the service they receive at the WSIB. The Commission also tracks complaint trends, identifies systemic issues and recommends improvements to the WSIB.
>
> Information about how to lodge a complaint with the Commission is provided on its website. A review of recent complaints indicates that most complaints are made by workers, and often relate to delays in the processing of claims by the WSIB or to concerns about service (for example, a number of complaints allege that claims adjudicators are overworked or hard to reach). Complaints by employers are fewer in number.
>
> One example of a complaint that led to a resolution in favour of an employer is summarized by the Commission on its website as follows:
>
>> An employer decided to cancel its voluntary WSIB insurance coverage after more than 27 years with no lost-time injuries. The WSIB applied a departure premium based on the employer's entry into the system in 1965.
>>
>> The employer was concerned with the lack of notice about the premium. ... The employer was also concerned about how the premium was calculated. ... The departure premium was almost double the employer's annual premium.
>>
>> The Commission made inquiries with the employer's account advisor, the revenue specialist, the assistant director of policy and the sector director. ... The director compared the cost of the premiums paid by the employer since 1965 to the employer's accident costs, concluded it was unfair to charge the departure premium, and cancelled it. (http://www.fairpractices.on.ca/html/cases/dmpemplydepart.htm)
>
> For the full text of this summary and to learn more about the Commission, visit its website.

WHAT HAPPENS WHEN A WORKER IS UNLIKELY TO RETURN?

In some cases, an injured worker does not return to his original workplace after recovery. There are a number of possible reasons for this:

- The employer may have employed fewer than 20 people, did not have a duty to hold the worker's job for him during a long recovery period, and does not have the need for a worker with the injured worker's capabilities.

- The worker may have worked for the employer for less than one year prior to the injury.

- The nature of the injured worker's post-injury limitations make him unable to do the kind of work he did prior to the injury.

- The worker requires substantial accommodation that would cause undue hardship to the original employer.

- The employer refuses to cooperate in promoting an ESRTW for the worker. The reality is that some accidents lead to a serious rupture of the relationship between worker and employer, and the parties may no longer wish to work together.

In some cases, the worker may simply apply for a different job and begin working somewhere else. However, where the worker is still recovering, struggling to find work, or still adjusting to a new disability, he may need extra help getting back on his feet. In these cases, the WSIB will make a determination that the worker is in need of and would benefit from a labour market re-entry plan.

A labour market re-entry plan can assist the worker either in returning to his original workplace or in finding a new position "in the employment or business that is suitable for the worker and is available" (WSIA, section 42(6)). The worker, his health care provider, and the WSIB will work together to determine what kinds of employment or business suit the worker's new circumstances, and will develop a plan to help prepare the worker to pursue appropriate opportunities. Often, a labour market re-entry plan incorporates retraining and job-search counselling, which will be paid for by the WSIB. The WSIB may also extend the period during which the worker receives LOE benefits to allow for completion of the plan.

Labour market re-entry plans are also offered, in certain circumstances, to the surviving spouses of workers killed on the job (for example, where the spouse did not work before the accident but must do so afterward in order to support children).

A worker is required to cooperate in all aspects of a labour market re-entry plan. Failure to complete the requirements of the plan can lead to the WSIB cutting off funding for the plan, or reducing/suspending other benefits.

REVIEW QUESTIONS

1. Is every employer required, by law, to provide work for a returning injured worker?

2. What role does an injured worker's health care provider play in paving the way for an ESRTW?

3. What is the purpose of the PDIF form, and who is responsible for filling it out?

4. Is the employer required to offer a returning worker the job he held before being away from work injured?

5. What is the duty to accommodate?

6. Is an employer required to provide every accommodation requested by the worker?

7. Why is good communication between the employer and the worker essential to a successful return to work?

8. What are the consequences, to the employer, of failing to support a worker's ESRTW?

CHAPTER 12
A Culture of Safety

CHAPTER OBJECTIVES

After completing this chapter, you should be able to:

- Understand what is meant by a "culture of safety."
- List the benefits of a culture of safety.
- Know how to assess the status quo, within an organization, with respect to safety attitudes.
- Explain the importance of influencing worker safety attitudes.
- Describe three ways in which employers can promote good safety behaviours.

INTRODUCTION

Safety as a facet of corporate culture has been touched on in other parts of this book, notably in chapter 7. That chapter, however, focused primarily on the relationships between worker safety and company productivity, and suggested that promoting worker safety is more likely than not to be cost effective for most employers. Chapter 7 also introduced the notion that some workplaces enjoy a "culture of safety" in which the employer and its workers share a commitment to safety; to this end, they cooperate in reducing risks—and thereby injuries—in the workplace.

One hallmark of highly successful companies is a commitment to excellence that permeates from the top ranks of management down to the most junior staff. Inspiring a personal commitment, on the part of workers, to furthering the interests of a corporation can sometimes pose a challenge. However, there exists an important opportunity for precisely this kind of alignment when it comes to occupational health and safety. When a company makes an explicit commitment to safety, workers recognize common ground between their own and the company's core values. This alignment, when backed up by safety strategies that workers can see, can encourage workers to align their priorities with the company's priorities on other fronts (for example, when it comes to solving productivity problems).

On its website at www2.dupont.com, DuPont Canada, a highly successful Canadian company, lists its core values in the following order: "Safety and Health;

Environmental Stewardship; Ethical Behavior; Respect for People," and has the following to say about its commitment to "reducing [its] footprint throughout the value chain":

> Safety, concern and care for people, protection of the environment, and personal and corporate integrity are this company's highest values and we will not compromise them.

This chapter explains what is meant by a culture of safety, and suggests strategies that may support the development of such a culture.

WHAT IS A CULTURE OF SAFETY, AND WHY STRIVE FOR ONE?

Organizational or corporate culture has been defined in many ways. Here are two sample definitions:

> [T]he specific collection of values and norms that are shared by people and groups in an organization and that control the way they interact with each other and with stakeholders outside the organization. Organizational values are beliefs and ideas about what kinds of goals members of an organization should pursue and ideas about the appropriate kinds or standards of behavior organizational members should use to achieve these goals. From organizational values develop organizational norms, guidelines or expectations that prescribe appropriate kinds of behavior by employees in particular situations and control the behavior of organizational members towards one another.

(Hill & Jones, 2001, p. 68.)

> The way we do things around here.

(Deal, 1993, p. 6.)

A well-developed corporate culture tends to influence management and workers alike to behave in certain ways. For example, some corporations pride themselves on being egalitarian, or "flatter" than average when it comes to the chain of command. In this kind of corporate culture, a free flow of ideas and influence might be expected to exist not only downward, but also laterally and upward within the authority structure.

In a safety culture, the answer to the question "How do things get done around here?" will always include the word "safely."

Embedding safety values into an organization's culture makes it more likely that safety will be given due consideration whenever managers make decisions and whenever workers take action. If findings in the field of cognitive psychology are correct, creating a culture of safety is one of the most cost-effective ways of reducing injuries, because incoming workers typically pick up quickly (even unwittingly) on workplace culture in an effort to fit in and to win recognition. A culture of safety has the potential to influence worker attitudes even with no special effort on the employer's part, and workers' adopted attitudes toward safety can be expected—at least most of the time—to guide their behaviour.

> ## Thinking Safe
> ### Leading with Safety
> While many employers strive to promote safety as one of their core values, a few companies have gone one step further: they have chosen to *lead* with safety—that is, to position worker safety as the most important value in the organization.
>
> One company that has had great success with this fairly radical approach is Alcoa, a leading US aluminum company. In 1987 Alcoa CEO Paul O'Neill tackled lagging corporate growth by making safety outcomes the *primary* indicator of senior leadership's performance. Between 1987 and 2002 Alcoa became a world safety leader, reducing its lost-time incident frequency rate from 1.86 to 0.12, per 100 workers per year, and increasing its market capitalization from $2.9 billion to $29.9 billion (a tenfold increase).

WHAT DOES A CULTURE OF SAFETY LOOK LIKE?

A culture of safety means that the way things get done in the organization is *safely*, and that all employees are responsible for reinforcing the employer's standards, expectations, and norms when it comes to safety. In practical terms, this means that management and employees at every level are expected to challenge incidents of unsafe behaviour (by confronting each other), to report problems, to refuse or stop unsafe work, and to reinforce the company's standards of safe work (for example, by asking others to wear protective devices or by teaching safe practices). It also means that these challenging, reporting, refusing, reinforcing, and teaching behaviours should be supported and rewarded by the company.

If employee work refusals are met with resistance or grumbling from supervisors, the company does not have a culture of safety. If supervisor requests for the wearing of PPE or the observance of safety procedures are met with resistance from workers, the company does not have a culture of safety. If workers' suggestions to senior management for solutions to safety problems are not thoughtfully considered and met with appropriate responses, the company does not have a culture of safety, *even if the rate of injuries is acceptably low.*

In an organization with a safety culture in place, safety-related requests and directives are accepted and followed with little or no conflict, because there is an underlying acceptance that "the way things get done around here is *safely.*" Within a culture of safety, managing safety is easy; without a supportive culture in place, managing safety is difficult. This is why safety performance must be managed from the top down.

ASSESSING THE STATUS QUO

Once a company has decided to foster a culture of safety, the first step it must take is an assessment of existing attitudes about safety on the part of both workers and management. *How* this assessment will be conducted will depend on the size of the company.

One step in assessing safety attitudes may be a review of internal complaint and discipline records. Have workers been complaining about safety? Have there been recent work refusals? If so, how were they handled? Are workers complaining about safety *rules* (which is very different from complaining about their own safety)?

What kinds of safety issues has management been struggling with? Have there been work refusals? Have there been worker requests for more training? Have workers been asking for safety improvements? Have they been complaining about safety procedures or equipment? Have workers been disciplined for failing to comply with safety rules? Is there a perception on the part of managers that certain safety policies limit productivity?

Once the records review is complete, the employer may wish to solicit input from employees. A questionnaire entitled "Personal Perceptions of Organizational Safety," developed by E. Scott Geller, is included in appendix J. This is one example of a tool that employers may use to gather information about existing safety attitudes. Using this questionnaire, or one of the employer's own creation, can help the employer gauge workers' perceptions of the safety culture—which may differ significantly from management's own perceptions.

Allowing a questionnaire to be completed anonymously may promote greater honesty in employee responses, but it also makes it impossible to make questionnaire completion mandatory. Having workers share their thoughts verbally, in a focus-group setting, is another option. Individual employers will have to decide how best to obtain input from their workers. It is important to keep in mind that to solicit input and then fail to act upon that input makes a negative impression. Employers serious about building a culture of safety should never ask for recommendations unless they are prepared to give them serious consideration.

Assessing the safety-attitudes status quo will allow the employer to identify immediate priorities, and will provide a benchmark against which it can compare its progress over time. Once the employer has an understanding of the status quo, it can move on to the task of influencing two things within the workplace:

1. safety beliefs; and
2. safety behaviours.

FACTS AND FIGURES
The Bystander Effect

One of the features of a culture of safety may be that employees are more likely to take personal responsibility for the safety of colleagues, which may help offset a psychological phenomenon known as "the bystander effect."

The phenomenon was first formally studied in 1968 by John Darley and Bibb Latane. The researchers confirmed the surprising fact that, the greater the number of bystanders available to help a victim (for example, an accident victim or a victim of violence), the less likely it was that the victim would be helped by anyone. In other words, where there is only one bystander, the victim is more likely to receive assistance than if there are many bystanders.

Psychologists have come to believe that, where there are multiple bystanders, the bystanders will rely on each other to help the victim instead of taking the responsibility upon themselves. Many reasons have been suggested for this, including the possibility that

- more bystanders means that each feels a "diffusion of responsibility" (that is, feels a shared, rather than an individual, responsibility);
- would-be helpers may be deterred from helping by a form of "stage fright"—shyness, or concern about "loss of face"; and

- individual bystanders may believe that other bystanders are *more qualified* to help.

Employers may be able to reduce the bystander effect by emphasizing the fact that safety is a key corporate value and is everyone's responsibility. Also key to minimizing the bystander effect is the provision of accident-response training: the more competent workers feel when it comes to responding to an incident, the more likely they will be to respond.

FOSTERING SAFETY BELIEFS

Introduction

The science of psychology is not a uniform code. Psychologists often have differing theories and beliefs about why people feel and behave the way they do. In examining psychological aspects of the working world, where employers are generally more focused on how people *behave*, there are two general streams of thought about how to influence worker behaviours.

Cognitive psychology suggests that, in order to change workers' behaviours, it is necessary to first influence how workers think and feel. In his book *Target Risk*, Gerald J. S. Wilde writes: "It's more important to reduce risk tolerance than increase compliance with specific safety rules" (p. 223).

Behavioural psychology suggests that it is more efficient to target behaviours directly, without becoming bogged down in worrying about influencing attitudes. A detailed discussion of psychology is beyond the scope of this text; however, in thinking about how to create a culture of safety in a workplace, it is probably useful to consider ways of influencing both worker beliefs and worker behaviours.

Risk Adversity, and Promoting Risk Intolerance

Central to worker safety beliefs is the idea of **risk adversity**. A person who is very risk adverse tends to take extreme steps to avoid risk, while a person who has very low risk adversity tends to behave recklessly. A position somewhere in the middle of this spectrum usually yields the best balance between safety and productivity.

risk adversity
the degree to which a person is motivated to avoid risk

However, even where a person has appropriate risk adversity, her beliefs about risk may not be in line with reality. For example, psychologists know that more people are afraid of snakes than are afraid of car travel; however, a person's risk of being killed in a car accident is much higher than her risk of dying from snakebite.

This kind of irrational assessment of risks can also happen in the workplace, especially where the worker

1. has done something (for example, used a machine with a safety guard off) so many times without being hurt that she seriously underestimates the statistical risk of eventually being hurt; or

2. does not know enough about the risks of an activity to develop an adequate sense of intolerance of those risks.

Workers who understand the risks in their workplace and who neither under- nor overestimate those risks are the most likely to take appropriate safety precautions. Assisting workers to understand the risks they face is the responsibility of the employer.

The employer can promote proper risk attitudes in several ways. It can

- provide specific training to explain the risks involved in specific tasks;

- without frightening workers unduly, provide accurate statistical information about workplace injuries;

- model appropriate risk adversity (for example, by requiring managers to wear hard hats and safety footwear while inspecting industrial work areas); and

- by communicating general messages that support safety behaviours (for example, in a change room where personal safety gear is stored, posting a poster that reads "Safety gear: Take the time to get it right").

Where supervisors believe that long experience with work tasks has dulled workers' perceptions of risks related to those tasks, it can help to occasionally draw workers' attention to the accidents that *do* occur. As long as it is not an invasion of other workers' privacy, one way of doing this is to personalize this information by providing information about victims. In chapter 5 of *Working Safe: How to Help People Actively Care for Health and Safety*, E. Scott Geller notes:

> It's human nature to downplay risks at work because, when we've escaped those risks over and over, we expect to escape them in the future. Creating a Total Safety Culture forces us to battle against this aspect of human nature. We need to increase people's perception of work-related risks if we hope to improve safety. One way to do this is to describe injuries in people terms, i.e., instead of saying that X activity presents a risk of X injury, relate the actual experience of the injury to a specific injured person with whom workers can empathize.

For example, at a team meeting, instead of simply reminding workers to keep their hands away from a part of a machine, a supervisor might report that, "Gurjit, a worker who has been employed at our sister plant in Hamilton for 11 years, suffered a deep cut to his palm last week while using this machine." This kind of anecdote serves at least four purposes:

1. It reminds workers that long experience with a machine or a procedure will not shield them completely from risks.

2. It may prompt them to visualize the accident, which can heighten risk adversity.

3. It may prompt them to identify with the victim (Gurjit), and to think about how a serious hand injury would affect their own lives ("How would I diaper the baby?"; "I'd have to give up squash, and right before the playoffs"; and so on).

4. It sends the message that the employer's safety directives are not simply arbitrary, but rather are motivated by experience and by a desire to prevent harm to other workers.

Building Worker Pride in the Organization's Safety Record

Another way to promote a general sense of **risk intolerance** is to encourage workers to take personal pride in working safely. A company that works hard to establish itself as a safety leader in the industry—particularly in a dangerous industry—will naturally have an advantage when it comes to recruiting workers. This kind of company can support its good safety record by letting workers know that they have been hired in part because of their safety attitudes. Workers like to be recognized: if they are encouraged to believe that their personal commitment to safety leadership has helped them get a job with a desirable employer, they will be more likely to "live up to" the employer's expectations when it comes to low risk tolerance. This is a wonderful example of the productive alignment of worker and employer interests that was discussed earlier in this chapter.

risk intolerance
the degree to which a person is unwilling to accept risk

HARD LESSON

Can Rewarding Good Safety Performance Backfire?

When taken too far, attempts to encourage worker support of a good safety record can backfire. There has been a trend—in some places and times in workplace safety history—to define workplace safety performance in terms of "injury-free days." This practice typically involves the counting of "reportable" injuries only (which may be defined by the employer and/or by safety legislation); in addition, a system of rewards (for example, worker bonuses or a points program that allows workers to earn gifts) may be instituted to mark the passage of injury-free milestones—for example, each injury-free month.

Unfortunately, there have been reported instances in which such a program has led workers to conceal injuries rather than report them in an attempt to avoid "letting the team down" or to earn rewards for themselves.

Thinking Safe

Tying Safety Performance to Career Success

One way in which a company can sustain *management* engagement with health and safety promotion is to make executive performance in supporting the safety culture and achieving safety goals reviewable on periodic performance reviews. Safety performance could also be tied to bonuses, and could be considered a prerequisite for promotion.

PROMOTING SAFE BEHAVIOURS

Introduction

While working to lower workers' risk tolerance involves the communication of general messages, promoting safe behaviours requires a great deal of specificity.

An employer cannot promote safe workplace behaviours effectively without first developing an excellent understanding of the risks in the workplace and the steps that can be taken to minimize those risks. Identifying and monitoring workplace risks is a process with which the employer will likely already be familiar; learning about ways to minimize those risks, however, may require some research.

Safety innovations and best practices are evolving and emerging all the time, and an employer seeking to develop a culture of safety must make a habit of seeking out new information and appropriate training on a consistent and regular basis. Workplace safety associations are often a very good source of information about safety developments. A list of some of these associations appears in appendix B.

Once an employer has identified a list of safety behaviours it would like to promote, a plan will need to be put into place to make the changes happen.

Information and Training

In some cases, employees will need nothing more than information about the expected behaviours, or perhaps simple training to ensure that they have a clear understanding of the instructions. This kind of information can be provided by the immediate supervisor, for example, at a team meeting.

Whether or not more formal training will be necessary will depend on certain factors, including:

- how radically the desired behaviour differs from existing behaviours;
- how complicated the desired behaviour/procedure is; and
- the message that the employer wishes to send about the importance of the change.

Sometimes, even where a new safety behaviour or procedure is relatively simple, the employer may choose to schedule a formal training session. Doing so—and taking time away from other work tasks to do so—will send a clear message that the change is important to the employer and that compliance is expected. Formal training prevents workers from being able to excuse any non-compliance by saying that they don't understand the new requirement, and provides the employer with an opportunity to introduce any rewards that may be offered to encourage compliance.

Reminders

Once training has been completed, it may be useful to support compliance through the use of reminders. Reminders can be verbal—for example, at the beginning of a shift, the foreman can check in with workers to ensure that they are wearing a personal protective item, such as safety shoes. In some large workplaces, reminders can be broadcast over an internal public announcement or internal television system. Finally, reminders can be written, for example, in the form of signs posted in the workplace.

The use of reminder signage is very common in the safety field. Signs reminding workers of a wide range of behaviours are commercially available from companies that specialize in precisely these products.

Where signs are used, sign placement is key. A reminder—sometimes called an "activator"—should be provided in a location as close as possible to where the behaviour must be done. For example, where slippery floors are a concern, no benefit will be achieved by posting a reminder to "Keep entryways clear of water and snow!" in the lunchroom. Instead, this sign should be placed where people entering the building will see it immediately. Likewise, signs reminding workers to wear PPE might be placed on the door that leads from the change room into the work area.

Reminder messages that encourage safety *attitudes* are typically very general in content (for example, a banner that reads "Think Safe!" and that runs along the bottom of an internal TV screen). These messages, if the employer wishes to use them, are often best placed where they can be easily associated with the company's identity. As an example, a sign on a company's front lawn might read:

ACME PAINTS AND SOLVENTS
An industry safety leader!

This message might also appear on employee pay stubs, or as a footer on memos.

Reminder messages that encourage particular safety *behaviours*, however, need to be much more specific. For example, in an area where there are pinch hazards, a sign that reads "Think Safe!" will be much less effective than one that reads "Where's Your Offhand?"

One of the challenges, with respect to reminder signage, is that people tend to become habituated to the familiar features of their environment quickly. They may notice—and think about—a safety message the first two or three times they see it, but after a relatively short time the sign will blend into the surroundings and the message may be ignored. For this reason, an employer will need to change safety signage regularly, which can become expensive. Thinking about creative ways to "freshen" safety messages can help save money; involving workers in the process may be an even better idea.

Consider the following ideas for creating fresh safety reminders:

- In an area readily visible to all workers (for example, above a sink), a whiteboard can be placed, with the workers' names listed in a column on the left. The person whose name appears first takes the first turn posting a safety message of her own creation in the column to the right of her name. The message stays up for a week, after which time the person whose name is next on the list erases his predecessor's message and adds his own.

- A set of small safety posters are put up one by one in an area where workers will see them. As the individual posters accumulate, these pieces of the "puzzle" form a larger image with its own message.

- A rectangular magnetic board is put up, with a box of magnetic letters and numerals placed beside it. Workers are invited to create coded "vanity plate"-style safety messages on the board for their colleagues to decipher.

Some behavioural psychologists suggest that signs that imply consequences are more effective than signs that are simply directive. This technique, however, must be used with caution: consider, for example, the insensitivity of the warning to "Tie back long hair or risk a scalping!" A more appropriate example is provided by E. Scott Geller in his book *Working Safe: How to Help People Actively Care for Health and Safety*, when he cites the text on bumper sticker, applied to a teenager's car, that reads: "How's My Driving? Call My Dad: 867-5309."

Feedback

Once employees have been trained in new safety behaviours and have had an opportunity to practise those behaviours in their work, it falls to supervisors to reinforce the behaviours through feedback.

Feedback can be negative or positive. While positive feedback is more supportive of worker morale and is usually the most appropriate form of feedback in safety contexts, supervisors who shy away from providing corrective feedback in the face of risky behaviour do a disservice to their workers. Where corrective feedback is necessary, it should be given in private, with an explanation of the importance of the desired behaviour and a reminder of the consequence (not the career consequences—though in some cases a reprimand may be necessary—but rather the potential consequences from injury to the worker and her colleagues). It is also a good idea to inform the worker that safety is a company priority.

The distinction between giving corrective feedback to counter risky behaviour and disciplining an employee for sustaining an injury is an important one for employers to make. In some workplaces, an injury-free record is held up as such an important goal that workers may come to view being hurt as somehow letting colleagues down. This is a counterproductive environment, and can lead to underreporting of accidents and near-misses. Corrective feedback should be reserved for situations in which there is an opportunity to *prevent* accidents.

While opportunities to give negative feedback arise naturally over time, opportunities for *positive* feedback may need to be created. To reinforce the fact that safety is both a team and an individual responsibility, positive feedback should be given in two forms:

- public and general praise to a work group or team; and
- private praise to individuals.

Offering intermittent praise for safety compliance increases the likelihood that workers will be receptive to corrective feedback, should such feedback ever be necessary.

Where appropriate, employers should also consider giving more formal praise, for example, by nominating an "employee of the month" or by highlighting a worker's safety achievement (for example, his suggestion for a modification to a workplace procedure) in a company newsletter.

As noted earlier in this chapter, companies with a culture of safety tie safety performance to career success (for example, by making it a factor in considerations for pay increases or promotions). Even where there is no formal program in place to measure and reward safety performance, it can be very useful when promoting an individual to remind him that his strong safety performance was a factor in his success in winning the promotion. This helps reinforce the message that safety is a priority at *all* levels in the organization.

Rewards

Where promoting a safety behaviour is of particular importance to an employer, the employer may decide to introduce a reward program to reinforce the new behaviour. In his article *Rate Your B.O.S.S.—Benchmarking Organizational Safety Strategy*,

Larry Hansen noted "What gets measured gets done; what gets measured and rewarded gets done well."

Because reward programs usually cost money (though it need not be a great deal of money) and require administrative time, most employers choose to run them for a limited time only. They may be instituted to coincide with the introduction of a new safety program or procedure, and may run for a few weeks; the "grand prize," if there is one, will be awarded at the end of a set period, by which time good compliance with the program is expected.

In order for reward programs to be—and appear—fair, the behaviour that is rewarded must either be readily measurable or winners must be nominated by their peers.

Since measuring safety performance can be challenging, employers will need to be creative in designing their programs. Consider, for example, a workplace where heavy articles need to be moved on a regular basis. Assume that, in response to a spate of back injuries, the employer has instituted a policy requiring two people to lift items that were previously lifted by one person. In this environment, each worker may be given a supply of small tokens (for example, poker chips). Any time a worker responds to a request by a colleague to help lift an object, the requesting worker must give the responding worker a chip; at the end of each shift, the workers can record the number of chips in their possession on a whiteboard, and at the end of the month the worker with the highest daily average of end-of-shift chips is named the winner and receives a prize. This kind of contest, if it appeals to the workers, will generally result in a spirit of "competitive volunteerism."

The employer may also choose to run a contest in which workers are asked to come up with ideas designed to increase safety in the workplace, with the best ideas warranting rewards.

Some psychologists suggest that a program in which rewards are won randomly and intermittently can be especially motivating. This kind of reward schedule—that is, one that is random and intermittent—is what makes slot machine gambling so addictive. An example of a random and intermittent reward program might involve workers being granted one raffle entry for every time they put on a new, unpopular piece of protective equipment at the beginning of a shift. At the end of each week, a draw is held, and the employees who have the largest number of entries will be more likely to win.

Rewards given through these programs need not be lavish. For example, the prize in a weekly draw might be a gift certificate for lunch at a local lunch spot, and the grand prize in a longer-term contest might be, for example, a personalized golf bag and a certificate for several golf lessons. E. Scott Geller notes that enhanced safety, not prizes, should be perceived as the major payoff in any incentive program (Geller, 1996).

KEY TERMS

risk adversity

risk intolerance

REVIEW QUESTIONS

1. What is organizational culture?
2. What is the benefit of an employer and its workers having aligned interests?
3. What is a culture of safety?
4. List three features of a culture of safety.
5. What is risk adversity? Risk intolerance?
6. List three ways employers can try to influence safety behaviours in the workplace.

APPENDIX A
Regulations Under the Ontario Occupational Health and Safety Act

GENERAL REGULATIONS

Control of Exposure to Biological or Chemical Agents, RRO 1990, reg. 833

Criteria to Be Used and Other Matters to Be Considered by Adjudicators Under Subsection 46(6) of Act, O. Reg. 243/95

Critical Injury—Defined, RRO 1990, reg. 834

Designations Under Clause 16(1)(N) of the Act, RRO 1990, reg. 847

Workplace Hazardous Materials Information System, RRO 1990, reg. 860

Training Programs, O. Reg. 780/94

Training Requirements for Certain Skill Sets and Trades, O. Reg. 572/99

Inventory of Agents or Combinations of Agents for the Purpose of Section 34 of the Act, RRO 1990, reg. 852

Joint Health and Safety Committees—Exemption from Requirements, O. Reg. 385/96

SECTOR-SPECIFIC REGULATIONS

Construction Projects, O. Reg. 213/91

Diving Operations, O. Reg. 629/94

Farming Operations, O. Reg. 414/05

Health Care and Residential Facilities, O. Reg. 67/93

Industrial Establishments, RRO 1990, reg. 851

Mines and Mining Plants, RRO 1990, reg. 854

Oil and Gas—Offshore, RRO 1990, reg. 855

Teachers, RRO 1990, reg. 857

University Academics and Teaching Assistants, RRO 1990, reg. 858

Window Cleaning, RRO 1990, reg. 859

HAZARD-SPECIFIC REGULATIONS

Designated Substances

Acrylonitrile, RRO 1990, reg. 835

Arsenic, RRO 1990, reg. 836

Asbestos, RRO 1990, reg. 837

Asbestos on Construction Projects and in Buildings and Repair Operations, O. Reg. 278/05

Benzene, RRO 1990, reg. 839

Coke Oven Emissions, RRO 1990, reg. 840

Ethylene Oxide, RRO 1990, reg. 841

Isocyanates, RRO 1990, reg. 842

Lead, RRO 1990, reg. 843

Mercury, RRO 1990, reg. 844

Silica, RRO 1990, reg. 845

Vinyl Chloride, RRO 1990, reg. 846

Other

Confined Spaces, O. Reg. 632/05

Needle Safety, O. Reg. 474/07 (comes into force September 1, 2008)

X-Ray Safety, RRO 1990, reg. 861

PROTECTIVE EQUIPMENT REGULATIONS

Roll-Over Protective Structures, RRO 1990, reg. 856

Firefighters—Protective Equipment, O. Reg. 714/94

APPENDIX B
General and Sector-Specific Health and Safety Associations

SAFE WORKPLACE ASSOCIATIONS

Construction Safety Association of Ontario (CSAO)
1-800-781-2726 or (416) 674-2726
Fax: (416) 674-8866
www.csao.org

Education Safety Association of Ontario
1-877-732-3726 or (416) 250-8005
Fax: (416) 250-9190
www.esao.on.ca

Electrical and Utilities Safety Association (EUSA)
1-800-263-5024 or (905) 625-0100
Fax: (905) 625-8998
www.eusa.on.ca

Farm Safety Association (FSA)
1-800-361-8855 or (519) 823-5600
Fax: (519) 823-8880
www.farmsafety.ca

Industrial Accident Prevention Association (IAPA)
1-800-406-4272 or (905) 614-4272
Fax: (905) 614-1414
www.iapa.ca

Mines and Aggregates Safety and Health Association (MASHA)
(705) 474-7233
Fax: (705) 472-5800
www.masha.on.ca

Municipal Health and Safety Association of Ontario (MHSAO)
(905) 890-2040
Fax: (905) 890-8010
www.mhsao.com

Ontario Forestry Safe Workplace Association (OFSWA)
(705) 474-7233
Fax: (705) 474-4530
www.ofswa.on.ca

Occupational Health and Safety Clinics for Ontario Workers (OHCOW)
1-877-817-0336
www.ohcow.on.ca

Ontario Safety Association for Community and Healthcare (OSACH)
1-877-250-7444 or (416) 250-7444
Fax: (416) 250-7484
www.osach.ca

Ontario Service Safety Alliance (OSSA)
1-888-478-6772 or (905) 602-0674
Fax: 1-877-250-9744 or (905) 602-6517
www.ossa.com

Pulp and Paper Health and Safety Association (PPSHA)
(705) 474-7233
Fax: (705) 472-8250 Confidential fax: (705) 474-0219
www.pphsa.on.ca

Technical Safety and Standards Authority
1-877-682-TSSA (8772) or (416) 734-3300
www.tssa.org

Transportation Health and Safety Association of Ontario (THSAO)
(416) 242-4771
Fax: (416) 242-4714
INWATTS 1-800-263-5016
www.thsao.on.ca

Workers Health and Safety Centre
Toll free: 1-888-869-7950
Central Ontario: phone (416) 441-1939, fax (416) 441-2277
Eastern Ontario: phone (613) 232-7866, fax (613) 232-3823
Northwestern Ontario: phone (807) 473-3634, fax (807) 473-3655
Northeastern Ontario: phone (705) 522-8200, fax (705) 522-8957
Southwestern Ontario: phone (519) 337-6083, fax (519) 337-6807
South-Central Ontario: phone (905) 545-5433, fax (905) 545-3131
www.whsc.on.ca

APPENDIX C
Guidelines for the Use of Professional Judgment in Classifying Controlled Products

i. Non-toxicological Criteria that Define the Limits for a Measurable Product Property when Subjected to a Specific Test Method which includes *CPR* sections 34 (*d*), 37, 38, 39 (*c*), 40, 65 (*a*):

A hierarchical approach to the consideration of test results should be used. The approach is described below.

1. Use test results on the product carried out in accordance with the specified test methods (either by conducting the test or using available test results). Professional judgement may be required to interpret results where, for example, there are varying test results for a product that has been subjected to the same specified test method.

2. In the absence of test results referred to in (1), use test results on the product from relevant but non-specified test methods. Professional judgement must be used with these results to classify the product.

3. In the absence of test results referred to in (1) or (2), where appropriate, extrapolate test results on a product with similar properties to classify the product. Professional judgement must be used to carry out such an extrapolation.

4. In the absence of being able to classify a product by steps (1), (2) or (3) above, a supplier must recognize that if the supplier sells the product and has classified it as not meeting the criterion and the product does in fact meet the criterion, the supplier will be in violation of the law.

ii. Toxicological Criteria that Define the Limits for a Measurable Product Property when Subjected to a Specific Test Method which includes *CPR* sections 46, 49, 52, 53, 55 (*b*), 57 (1) (*b*), 59, 60, 61 (*a*), 62, 65 (*b*) 66 (*c*):

A supplier must use a hierarchical approach to the consideration of test results as shown in steps (1) to (4) below *or* the approach shown in (5) below.

1. Use test results on the product carried out in accordance with the specified test methods (either by conducting the test or using available test results). Professional judgement may be required to interpret results where, for example, there are varying test results for a product that has been subjected to the same specified test method.

2. In the absence of test results referred to in (1), use test results on the product from relevant but non-specified test methods. Professional judgement must be used with these results to classify the product. Examples of relevant test methods are given in *CPR* paragraph 33 (3) (*b*).

3. In the absence of test results referred to in (1) or (2), where appropriate, extrapolate test results on a product with similar properties to classify the product. Professional judgement must be used to carry out such an extrapolation. (Although suppliers are not obliged, they are encouraged to make use of available quantitative structure-activity relationship (QSAR) systems to estimate the toxic effects of chemicals. Professional judgement is required to assess the value of such estimates.)

4. For Class D criteria, if it is not possible to classify a product by steps (1), (2) or (3), a supplier is not required to undertake toxicological testing. The product can be considered as not meeting a Class D criterion, if the supplier has no "information of which the supplier is aware or ought reasonably to be aware". Every supplier "ought reasonably to be aware" of appropriate published literature. The Canadian Centre for Occupational Health and Safety (CCOHS) is one organization capable of conducting a comprehensive literature search. When additional information is made available to the supplier by the appropriate regulatory agencies, by industry or trade association(s), and by labour organization(s), the supplier is expected to evaluate that information.

5. For Class D criteria, a supplier may use an alternate strategy *in place of* steps (1) to (4) above. A supplier may undertake a search of information he "ought reasonably to be aware of" (as in (4) above). If the supplier finds "sufficient" human data to show that the product meets or does not meet a criterion, the supplier may use this information to classify the product. Professional judgement must be used in making an assessment of what is "sufficient" in each case and taking into account animal test results.

iii. Criteria that Define the Limits for a Measurable Property or Qualitative Characteristic of a Product without Specifying a Test Method which includes *CPR* sections 34 (*a*) (*b*) (*c*), 36, 39 (*a*) (*b*), 41, 42, 66 (*a*) (*b*) (*c*):

None of these *CPR* criteria relate to Class D and therefore the supplier is obliged to use the direction provided in *CPR* subsection 33 (1) (*b*).

For the above criteria other than sections 39 (*b*) and 66 (*a*) (*c*), the use of professional judgement in classification is addressed in I(2) and (3) above.

For sections 39 (*b*) and 66 (*a*) (*c*), it is clear the professional judgement must be used to decide if the qualitative criteria properly describe the properties of the product.

iv. Criteria which State that "There is Evidence" of a Physiological Effect, without Specifying a Test Method which includes *CPR* sections 55 (*a*), 56, 57 (1) (*a*), 61 (*b*), 64, 65 (*e*):

The supplier must use professional judgement to decide if test results or studies on the product signify "evidence" of an effect. This would include giving consideration to the particulars of the test method or study and the relevance of the results or conclusions in the occupational situation. There is nothing in the *CPR* to prevent a supplier from over-classifying a product.

Where the supplier finds "evidence" that the product meets a criterion and also finds "evidence" to the contrary, the supplier must consider the product as meeting the criterion for the purpose of classification. The supplier may make reference to the contrary evidence on the MSDS, but such disclosure must be done in accordance with the qualifications referred to in section 13 of the *CPR*.

Where a supplier cannot find test results, conclusions from a study or other evidence on the product for one of these criteria, the supplier is not required to test the product but may assume, for the purposes of classification, that the product does not meet that criterion.

v. Criteria for Carcinogenicity—*CPR* section 54

There is no opportunity to use professional judgement in the classification of carcinogens when the substance or tested mixture is included in the referenced lists. The WHMIS criteria for carcinogens apply only to products or substances, not to processes listed by IARC or ACGIH, such as "antimony trioxide production" or "manufacture of magenta."

Where a substance or tested mixture does not appear on the referenced lists and the supplier has information to show that the product may be a carcinogen, the supplier should use professional judgement to decide if the product should be classified as carcinogenic. While it is required that such information be disclosed on the MSDS of a product, a classification of the product as carcinogenic would not be required by WHMIS legislation.

vi. Criteria that Refer to *Transportation of Dangerous Goods Regulations (TDGR)* Criteria which includes *CPR* sections 39 (*d*), 47, 50, 65 (*c*) (*d*):

The referenced TDG criteria are of the type referred to in III above or, in the case of section 65, of the type referred to in IV above. The same appropriate rules for the use of professional judgement apply to these criteria.

In addition to containing scientific criteria, the TDGR contains a list of specified dangerous goods in Schedule II with designated primary classifications. If a product is listed in Schedule II to the *TDGR* as meeting one of the referenced TDG criterion, the supplier cannot use professional judgement to decide that the product does not meet that criterion.

Suppliers should be cautious when reading the TDG classification in Schedule II. TDG prioritizes the hazards and only lists the most severe hazard in Schedule II. Thus, when assessing a product against a particular TDG criterion, a supplier should first refer to the Schedule II list and, where the product is not listed as meeting the criterion, also refer to the TDG criterion in the TDG Regulations before concluding the product does not meet the criterion.

vii. Criteria for Ingredients in a Product that is an Untested Mixture which includes *CPR* sections 48, 51, 58, 63, 65 (*f*):

The same rules for use of professional judgement that applied when deciding if a tested product meets a criterion will apply when deciding if an ingredient is a controlled product.

Note: The WHMIS legislation does not prohibit a supplier from including a product in any WHMIS class, division or subdivision even though it does not strictly meet the hazard criteria in the CPR. However, suppliers should avoid including products that are clearly beyond the scope of the hazard criteria that define a class. Otherwise, the overall effectiveness of WHMIS in accurately warning workers of the hazards inherent in workplace products will be diminished.

If a supplier's product falls just outside the criteria which define any class, a supplier may use professional judgement to decide that the product should nonetheless be included in the class.

Source: Health Canada, Environmental Health & Workplace Health, Occupational Health & Safety, Workplace Hazardous Materials Information System, Classification, http://www.hc-sc.gc.ca/ewh-semt/occup-travail/whmis-simdut/classification_e.html. Reproduced with permission of the Minister of Public Works and Government Services Canada, 2008.

APPENDIX D

Sample Material Safety Data Sheet (MSDS) for Silica Sand

MATERIAL SAFETY DATA SHEET

ATLANTIC SILICA INC.

Section 1	MATERIAL IDENTIFICATION AND USE	
Material Name / Identifier: Silica sand or Quartz sand		
Supplier: Atlantic Silica Inc. 4 Osborne Road Poodiac, New Brunswick, E4E 5K5, Canada		**Information Telephone #'s** (506)-433-5890 or (902)-883-3020
Chemical Name Silicon dioxide		**Chemical Family** Inorganic oxides
Intended Material Uses Typical uses of this product include but are not limited to sand use for: beach renewal, cement additives, concrete aggregate, construction sands, decorative sands, fertilizer filler, fiberglass, filtration sands e.g. pools, golf course sands, lawn bowling sands, mortar mixes, playground sands, mixing for sheet and container glass, silicon carbide production, top dressing sand, tractions sands, tree nursery grits and white golf bunker sand. **Caution for Other Uses** Using this product as blasting sands, foundry sands, refractory sands, smelter flux, can/will result in the formation of additional airborne and respirable quartz/silica dust and crystals, which are smaller in size than the original product. Such subsequent uses therefore have the primary potential for harm to workers, described in this MSDS document, due to the ability of the smaller particles to penetrate further into the lung.		**Trade Names & Synonyms** #00 Sand, #0 Sand, #1 Sand, #2 Sand
Regulatory Classification **WHMIS:** Class D2B, Poisonous and Infectious Materials Causing Other Toxic Effects, Chronic Toxic Effects **NFPA Ratings:** No rating **TDG:** Not required, not a Transportation Dangerous Good		

Section 2		HAZARDOUS INGREDIENTS		
Hazardous Ingredients	Approx. Conc. % (#00 Sand)	C.A.S. Number	Exposure Limits (ACGIH 2007)	LD50 / LC50 Specify Species and Route
Silica Crystalline Quartz	30-60%	1480-60-7	TLV = 0.025 mg/m^3 as respirable quartz particle size less than 4 um	Not available
Inhalable particles (PNOS)	20%	Not applicable	TLV = 10 mg/m^3 particle size less than 100 um	Not available
"Respirable particles"	0.3% (or less)	Not applicable	TLV = 3 mg/m^3 particle size less than 4 um	Not available.

Section 3	PHYSICAL DATA		
Physical State Solid	**Odor and Appearance** Odourless fine grained particles white to yellowish white		**Odor Threshold (ppm)** Not applicable
Specific Gravity (H$_2$O=1) 2.66	**Vapor pressure** (mm Hg & temp) 0 mm Hg	**Vapor Density** (Air =1) None	**Evaporation Rate** Not applicable
Boiling Point (°C) 2230	**Freezing Point (°C)** 1710	**pH** Not applicable	**Coef. of Water/Oil Dist.** Not applicable
Solubility in Water Insoluble	**Solvent Solubility** Insoluble in common solvents. Soluble in hydrofluoric acid.		

Atlantic Silica Inc.: MSDS: Silica Sand: Revision Date, June 27, 2007

MATERIAL SAFETY DATA SHEET

Section 4 — FIRE AND EXPLOSION HAZARDS

Flammability – If yes, under which conditions:		
Not flammable		

Means of Extinction –		
None required, material not flammable. Sand is used as means to extinguish Class A and B fires.		

Flash Point (°C) and Method	Upper Explosion Limit	Lower Explosion Limit
Not applicable	Not applicable	Not applicable
Auto Ignition Temperature (°C)	**Hazardous Combustion Products**	**Explosion Data / Sensitivity to Mechanical Impact**
Not applicable	Not applicable	None
Rate of Burning	**Explosive Power**	**Sensitivity to Static Discharge**
Not applicable	Not applicable	None

Section 5 — REACTIVITY DATA

Chemical Stability – If no, under which conditions?
Substance is extremely stable.

Incompatibility with other substances. If so, which ones?
Powerful oxidizers: fluorine, chlorine trifluoride, manganese trioxide, oxygen difluoride, hydrogen peroxide, etc. Acetylene or ammonia.

Reactivity and under what conditions?
With hydrofluoric acid to produce a corrosive gas silicon tetrafluoride.

Hazardous Decomposition Products:
Under mechanical impact (e.g. blasting) silica sand physical decomposes to fine dust or respirable silica quartz (particles less than 4 microns diameter).

Hazardous Polymerization? Will not occur.

Section 6 — TOXICOLOGICAL PROPERTIES

Prime Route of Entry   Inhalation of airborne respirable particles/silica.   **Secondary** None

Effects of Acute Exposure:
Acute hypersensitivity reactions or rapid developing silicosis may occur in a short period of time in heavy unprotected exposure situations, such as sandblasting and foundry use. Particles in levels above the TLV may cause shortness of breath, dryness of mucous membranes, and irritation to eye, nose and throat.

Effects of Chronic Exposure:
1. Prolonged exposure can cause silicosis, pulmonary fibrosis, other obstructive pulmonary conditions and cancer of the lung. Quartz is listed by ACGIH as an A2, a suspected human carcinogen.
2. Scleroderma or Progressive Systemic Sclerosis (PSS) an Auto-Immune disease affecting connective tissues and many organ systems of the body.
3. Possible kidney damage (chronic nephropathy) with resulting reduced kidney function.

LC50	Irritancy	Exposure Limits
Not available	Airborne particles will irritate unprotected eyes and other mucous membranes.	0.025 mg/m^3 as respirable silica quartz < (less than) 4 micron particle size
LD50	**Sensitization**	**Synergistic Materials**
Not available	None known	None known

Atlantic Silica Inc.: MSDS: Silica Sand: Revision Date, June 27, 2007

MATERIAL SAFETY DATA SHEET

PAGE 3 OF 4

Toxic Effects: Component, respirable quartz
■ Carcinogenicity ☐ Mutagenicity ☐ Reproductive Toxicity ☐ Teratogenicity

General: Silicosis is a form of disabling pulmonary fibrosis which can be progressive, adversely affect lung function and may lead to death. For workers having chronic exposure (frequent, regular, long term) Baseline and ongoing medical monitoring is recommended. OSHA recommends that individuals chronically exposed to silica receive baseline and periodic medical and lung function examinations. Such examinations should be repeated every five years in individuals with less than 20 years exposure and every two years in those with more than 20 years exposure. A chest XRAY is recommended upon termination of employment. As well, where respirable silica levels exceed exposure limits it is prudent to implement a mandatory respiratory protection program.

Section 7 — PREVENTATIVE MEASURES

Personal Protective Equipment
Preventing Silica Crystals from entering the lungs is imperative to prevent acute or chronic harm. Develop and enforce a respiratory protection program to minimize harm to workers exposed to respirable silica.

For sand blasting use only NIOSH approved Type CE, Abrasive Blast Supplied-Air Respirators. Depending on measured or known exposure levels use continuous-flow respirator loose fitting hood for dust levels up to 25 X the TLV, continuous flow tight fitting face piece to 50 X the TLV, positive-pressure respirator with tight fitting half-mask face piece to 1000 X the TLV. See users note at www.cdc.gov/niosh/abrasi-1.html dated May 23, 1996.

Gloves (Specify)	Respiratory (Specify)
As necessary for physical protection from blasting or foundry, etc use. **Note:** Severe injury can occur when in contact with sand stream under pressure during sand blasting.	For non application use adjacent workers should wear NIOSH approved N100, R100, P100 or HEPA half face dust masks up to 10 X the TLV, or as a precaution against accidental exposure. R100 filters should be replaced once 200 mg total dust (not just quartz) is on the dust mask. This use time can be estimated by knowing average dust levels (TWA) in mg/m^3 x inhaled air/shift (average = approx. 10 m^3 shift).
Eye (Specify)	**Footwear (Specify)**
As necessary for physical protection from blasting or foundry, etc use.	As necessary for physical protection from blasting or foundry, etc use.
Clothing (Specify)	**Other (Specify)**
As necessary for physical protection from blasting or foundry, etc use.	It is recommended the user have a Respiratory Protection Program in accordance with CSA Z94.4-02 Selection, Use, and Care of Respirators, or ASTM E 1132-86 Standard Practice for Health Requirements Relating to Occupational Exposure to Quartz Dust.

Engineering Controls (Specify)
Local exhaust with sufficient supply air and full exhausting of the occupied space (if enclosed or partly enclosed) sufficient to reduce levels below ½ protective capability of the chosen respirator, or if no respirator specified, below ½ the TLV. ½ TLV is specified due to the generally variable nature of exposure to dusts that could cause excursions over the TLV. See also ACGIH Industrial Ventilation, A Manual of Recommended Practice, latest edition.

Leak and Spill Procedures
For unused material collect using methods that do not raise dust levels, such as vacuum or shovel. Wear respiratory protection if sweeping is necessary. For used material, high levels of respirable silica quartz may be present do not sweep, use vacuums with HEPA filters or wet down with water before shoveling/ sweeping.

Waste Disposal
Unused material can be reused or disposed to landfill directly. Used silica quartz sand must be tested for contaminants in accordance with Federal, Provincial/State and Municipal regulations and disposed of in accordance with the authority having jurisdiction. This generally refers to bulk and/or leachate analysis for contaminating toxic metals (such as lead in removed paint).

Handling Procedures and Equipment
Avoid creating dust or mechanical abrasion of the material.

Storage Requirements
No special requirements, store covered and away from incompatible reactive compounds.

Special Shipping Information
Material has no TDG labeling or manifesting requirements.

Atlantic Silica Inc.: MSDS: Silica Sand: Revision Date, June 27, 2007

MATERIAL SAFETY DATA SHEET

Section 8	FIRST AID MEASURES
Skin	No hazard on skin, unless applied under pressure (e.g. sand blasting); may cause itching. ***Abrasions*** – Rinse abrasion to remove sand and seek medical attention. ***Burns*** – Treat as for abrasions.
Eye	Irrigate eye with water to remove sand and seek medical attention if irritation continues.
Inhalation	For acute inhalation remove person to fresh air. Seek medical attention if worker experiences difficulty breathing after exposure.
Ingestion	If material ingested and is not used material containing contaminants, make comfortable and provide water. If material is used and may contain contaminants, make person comfortable and provide water, and seek medical attention.
General Advice	Smoking and/or exposure to other respiratory irritants or toxins aggravate the effects of exposure.

Section 9	PREPARATION OF M.S.D.S.

Additional Information and Caution
If this product is used for other than intended uses, aggressive respiratory protection measures, as referenced above, are necessary to prevent the harm to exposed workers. See Section 1, "Intended Material Uses" for information on intended uses and cautions related to other uses.

Sources Used
MSDS Atlantic Silica Sand 2004, ACGIH 2007 TLVs and BEIs, NIOSH Pocket Guide to Chemical Hazards Feb 2004, NIOSH respirators user notice Type CE Respirators May 23, 1996, quartz analysis April 2004, chemical and physical analysis data Gallant Aggregates Limited Jan - Mar 2004, physical & size analysis (Maxxam Analytics Inc.) May 2007 #00 Sand.
Toxicology References: Clinical Environmental Health and Toxic Exposures, 2nd Ed., J. B. Sullivan and G. R. Krieger, Editors, 2001; and Casarett & Doull's Toxicology, 6th Ed., C. D. Klassen, 2001; OSHA Final PEL Rule (1989) and NIOSH Hazard Review of Silica (General), 2002.

Prepared by Clive MacGregor, M.Sc., ROH, MacGregor and Associates; and Greg Johnstone M.Sc., COHSM, PharmaTox Inc., Halifax, Nova Scotia, Canada. Prepared in accordance with Schedule I of the Controlled Products Regulations (Section 12) Information to be disclosed on a Material Safety Data Sheet (MSDS).

Checked by
John Cooper, Safety Advisor, Basin Contracting Limited (Affiliated Company of Atlantic Silica Inc.)

Preparers Phone Number	**Preparation Date** June 27, 2007
(902)-499-0320 (Clive MacGregor)	(Replaces Version July 2004)
(902)-468-1095 (Greg Johnstone)	

CAUTION:

The information contained herein is based on the information available at the indicated date of preparation but no warranty, expressed or implied, is made. Further, the information contained herein relates only to this product or material and may not be valid when used in combination with any other product or material in any process. If the product is not to be used for a purpose or under conditions that are normal or reasonably foreseeable, this information cannot be relied upon as complete or applicable. For greater certainty of information, specific uses of the product must be reviewed with the supplier.

Atlantic Silica Inc.: MSDS: Silica Sand: Revision Date, June 27, 2007

Source: Atlantic Silica Inc., http://www.atlanticsilica.ca/images/msdsjun27_07.pdf.

APPENDIX E

Sample Form for Use in Creating a Material Safety Data Sheet

SECTION 1 – CHEMICAL PRODUCT AND COMPANY IDENTIFICATION

Product Identifier			[WHMIS Classification]	
Product Use				
Manufacturer's Name		Supplier's Name		
Street Address		Street Address		
City	Province	City		Province
Postal Code	Emergency Telephone	Postal Code		Emergency Telephone
Date MSDS Prepared	MSDS Prepared by	Phone Number		

SECTION 2 – COMPOSITION/INFORMATION ON INGREDIENTS

Hazardous Ingredients (specific)	%	CAS Number	LD_{50} of Ingredient (specify species and route)	LC_{50} of Ingredient (specify species)

SECTION 3 – HAZARDS IDENTIFICATION

Route of Entry: ☐ Skin Contact ☐ Skin Absorption ☐ Eye Contact ☐ Inhalation ☐ Ingestion

[Emergency Overview]

[WHMIS Symbols]

[Potential Health Effects]

SECTION 4 – FIRST AID MEASURES

Skin Contact

Eye Contact

Inhalation

Ingestion

Sample 16 Section MSDS Format *[Optional, not required under WHMIS]* *Page 1 of 4*

Product Identifier

SECTION 5 – FIRE FIGHTING MEASURES

Flammability ☐ Yes ☐ No	If yes, under which conditions?
Means of Extinction	

Flashpoint (°C) and Method	Upper Flammable Limit *(% by volume)*	Lower Flammable Limit *(% by volume)*
Autoignition Temperature (°C)	Explosion Data – Sensitivity to Impact	
Hazardous Combustion Products		
[NFPA]		

SECTION 6 – ACCIDENTAL RELEASE MEASURES

Leak and Spill Procedures

SECTION 7 – HANDLING AND STORAGE

Handling Procedures and Equipment
Storage Requirements

SECTION 8 – EXPOSURE CONTROL/PERSONAL PROTECTION

Exposure Limits   ☐ ACGIH TLV   ☐ OSHA PEL   ☐ Other (specify)
Specific Engineering Controls *(such as ventilation, enclosed process)*
Personal Protective Equipment   ☐ Gloves ☐ Respirator ☐ Eye ☐ Footwear ☐ Clothing ☐ Other
If checked, specify type

Sample 16 Section MSDS Format   [Optional, not required under WHMIS]

Product Identifier

SECTION 9 – PHYSICAL AND CHEMICAL PROPERTIES

Physical State	Odour and Appearance	Odour Threshold (ppm)
Specific Gravity	Vapour Density (air = 1)	Vapour Pressure (mmHg)
Evaporation Rate	Boiling Point (°C)	Freezing Point (°C)
pH	Coefficient of Water/Oil Distribution	[Solubility in Water)

SECTION 10 – STABILITY AND REACTIVITY

Chemical Stability ☐ Yes ☐ No

Incompatibility with Other Substances ☐ Yes ☐ No

Reactivity, and under what conditions?

Hazardous Decomposition Products

SECTION 11 – TOXICOLOGICAL INFORMATION

Effects of Acute Exposure

Effects of Chronic Exposure

Irritancy of Product

Skin Sensitization	Respiratory Sensitization
Carcinogenicity - IARC	Carcinogenicity – ACGIH
Reproductive Toxicity	Teratogenicity
Embryotoxicity	Mutagenicity
Name of Synergistic Products/Effects	

Sample 16 Section MSDS Format   *[Optional, not required under WHMIS]*

SECTION 12 – ECOLOGICAL INFORMATION

[Aquatic Toxicity]

SECTION 13 – DISPOSAL CONSIDERATIONS

Waste Disposal

SECTION 14 – TRANSPORT INFORMATION

Special Shipping Information

		PIN
TDG	[DOT]	
[IMO]	[ICAO]	

SECTION 15 – REGULATORY INFORMATION

[WHMIS CLASSIFICATION]	[OSHA]
[SERA]	[TSCA]

This product has been classified in accordance with the hazard criteria of the Controlled Products Regulations (CPR) and the MSDS contains all of the information required by CPR.

SECTION 16 – OTHER INFORMATION

Sample 16 Section MSDS Format   [Optional, not required under WHMIS]

Source: Saskatchewan Labour, Occupational Health and Safety Division, http://www.labour.gov.sk.ca/MSDS16.

APPENDIX F
Costing an Individual Incident/Injury Form—from the WSIB

Costing an Individual Incident/Injury

Name: _____

Date and Time of Injury/Incident _____

Location: _____

Brief Description of Accident/Injury _____
_____
_____
_____
_____

Injury/Incident Costing

Direct Injury Costs	COST	
Compensation/Medical/Rehabilitation Costs Description of Injury: _____ Time Away from Work: _____ days/weeks		Section A
1. Compensation/medical/rehabilitation costs. (Use the "Estimated Actual Claim cost" from your organization's latest NEER statement. If the claim has had the Maximum Claim cap applied, use the value before the cap has been applied.)	$_____	
2. Wage supplements/Continuation of benefits etc	$_____	
3. Other compensation/Benefit Related Costs	$_____	
Total Direct Injury Costs	$_____	
Indirect Accident Costs	**COST**	
Property/Equipment/Material Damage Was there equipment or structural damage? Was it repaired or replaced? Was equipment rented for temporary replacement? Was there material or product damage or loss? Summary: _____ _____		Section B
1. Costs of Cleanup/salvage	$_____	
2. Equipment Repair/Replacement Cost	$_____	
3. Property/Structural Repair Cost	$_____	
4. Equipment Rental Cost	$_____	
5. Material/Product Loss	$_____	
6. Material/Product Rework	$_____	

Administrative Costs

Administrative costs include extra management and administrative time as a result of the injury/illness. Examples include managing the accident scene, conducting accident investigation and follow up actions, arranging compliance to MOL orders, rescheduling of people and work, arranging / supervising clean up and salvage, arranging for repair or replacement of equipment or other property, hiring replacement staff, training of replacement staff, follow up with injured or ill employee, arranging for return to work and modified work.

1. Management effort at time of incident/injury $_____
2. Management effort: clean up, restoration, rescheduling, MOL order compliance, replacement hiring and training etc. $_____
3. Follow up with injured employee including return to work, modified work etc. $_____
4. Administrative effort in claims management, medical and personnel effort, public relations etc. $_____
5. Lost opportunity cost: If management had not been spending its time on incident follow up, what could they have been doing? What is the value of that lost opportunity? $_____

Section C

Productivity Costs

These involve the costs associated with the work interruption due to the incident, and ability to carry out work following the incident. Calculations for these cost items will vary depending on how your organization values the cost of work interruptions, equipment unavailability, and reduced employee capability. There is also the lost opportunity cost, as above.

1. Work interruption at time of incident (# of people, length of interruption) $_____
2. How was work time made up: costs, overtime etc $_____
3. Rescheduling Costs: New set up, delays in equipment availability, idle people, time, costs $_____
4. Cost of learning curve of replacement employees $_____
5. Equipment out of service: MOL stop work orders, time to repair damage or replace. $_____
6. What was the impact of loss of use of equipment, processes, or skills of injured person $_____
    - Alternate processes more costly, time consuming
    - Customer Service issues: out of stocks, delays in delivery
    - Scheduling inefficiencies
7. Cost of reduced productivity while employee on modified work. $_____

Section D

MOL Orders

1. Cost of complying with MOL orders $_____
    - Equipment/process modifications
    - Administrative Procedures
2. Consultant Fees $_____

Section E

Legal Costs	
1. Legal Counsel Fees	$_____
2. Management/Employee time (preparation, witnesses, etc)	$_____
3. Expert Witness Fees	$_____
4. Fines	$_____
5. Other Costs (Settlements etc)	
Other Costs	
1. _____	$_____
2. _____	$_____
3. _____	$_____
Total	$_____

Instructions to Complete Costing an Individual Accident/Injury Form
To determine Indirect Accident/Injury costs using this form, ADD up Sections B, C, D, E, F, and G. Do NOT use Section A in this calculation. Do this process for all Accidents/Injuries for a given year. Use the total for ALL accidents/injuries in a given year in your calculation for **LINE B** on the **Business Case for Health & Safety Worksheet** on page 19 of this document.

Source: Workplace Safety and Insurance Board, "Business Results Through Health and Safety," appendix D, http://www.wsib.on.ca/wsib/wsibsite.nsf/Public/ReferenceEmployers.

APPENDIX G

Resources for Safety Managers, Health and Safety Representatives, and JHSC Members

Note: This is a sample listing only. Professionals with responsibility for occupational health and safety may be able to identify many more available resources.

PUBLICATIONS

Canadian Centre for Occupational Health and Safety (CCOHS), *Pandemic Planning*; available at www.ccohs.ca/pandemic.

Canadian Human Rights Commission, *A Guide for Managing the Return to Work* (2007).

Cientia.net provides an extensive list of occupational health and safety guides and other resources at: www.cientia.net/Cientia.Safety.Learning.htm.

Industrial Accident Prevention Association (IAPA), *Determining Significant Hazards at Work, A Guide for Employers and JHSCs* (March 2006); available at www.iapa.ca.

Institut de recherche Robert-Sauvé en santé et en sécurité du travail (IRSST) provides a list of guides for the management of specific safety challenges—some in French—for employers, see www.irsst.qc.ca.

Krause, Thomas R., *Leading with Safety* (John Wiley & Sons, 2005).

Ontario Bar Association, "Occupational Health and Safety Prosecutions: What You Need to Know" (seminar materials binder) from seminar held December 10, 2007; available at: www.softconference.com/oba/eventdetails.aspx?userID=24126624966175930200794300&isCLE=yes&code=07LAB1210C.

Ontario Ministry of Labour, *A Guide for Joint Health and Safety Committees (JHSCs) and Representatives in the Workplace* (revised June 2002).

Ontario Ministry of Labour, *A Guide for Health and Safety Representatives and Joint Health and Safety Committees on Farming Operations* (2005).

Ontario Ministry of Labour, *A Guide to the Occupational Health and Safety Act* (revised June 2002).

Workplace Safety and Insurance Board (WSIB), "Appeal System Practice and Procedures" (revised annually); available at www.wsib.on.ca.

ASSESSMENT TOOLS

The Canadian Centre for Occupational Health and Safety (CCOHS) provides health and safety inspection checklists appropriate to different kinds of workplaces (for example, manufacturing, offices). See www.ccohs.ca/oshanswers/hsprograms.

Industry-specific inspection checklists are often available from industry-specific health and safety associations (for example, the Transportation Health and Safety Association of Ontario (THSAO)). For a list of some of these associations, see appendix B.

Industrial Accident Prevention Association (IAPA), *Health and Safety Action Plan Form* (October 2007); available at www.iapa.ca.

Institut de recherche Robert-Sauvé en santé et en sécurité du travail (IRSST) provides a list of online assessment tools for both workers and employers, see www.irsst.qc.ca.

WSIB, "2008 Net Average Earnings Calculator," available at: http://www.wsib.on.ca/wsib/wsibsite.nsf/public/ReferenceEmployers.

WSIB, "Compliance Audit for Employers," available at: http://www.wsib.on.ca/wsib/wsibsite.nsf/public/ReferenceComplianceAuditEmployers.

WSIB, "Map Cost Tool," available at http://www.wsib.on.ca/wsib/wsibsite.nsf/public/ReferenceEmployers.

Workwell, a WSIB program, offers a "Core Health and Safety Audit," available for download at: http://www.wsib.on.ca/wsib/wsibsite.nsf/public/ReferenceEmployers.

TRAINING RESOURCES AND SAFETY PROGRAM SUPPLIES

Bright Training and Safety Wear

Bright Training and Safety Wear is a Hamilton, Ontario-based provider of training courses for various industrial skills, such as forklift driver operator training, overhead crane operator training, lockout tagout procedures, WHMIS training, and propane safety. The company also offers general OHS training in the areas of accident inspections, supervisory skills, corporate strategic planning, WSIB claims management, and Workwell audit compliance, among others. Visit Bright Training and Safety Wear's website at www.bright-training-safety-wear.com (contact information for local consultant partners listed online).

The Canadian Centre for Occupational Health and Safety (CCOHS)

The CCOHS offers a course called "Health and Safety Committees"; to register, see the CCOHS's annual course catalogue. The CCOHS also offers a long list of hazard- or issue-specific courses useful to committee members seeking to manage specific hazards; see, for example, "Ladder Safety" and "Personal Protective Equipment." CCOHS Webinars (internet-based seminars) are also offered on a range of topics, including "Pandemic Planning" and "Workplace Violence Awareness." Visit the CCOHS's website at www.ccohs.ca.

Dunk & Associates

These OHS consultants offer training and other OHS services. Visit them at www.dunk-andassociates.com.

Safety Smart

Safety Smart offers prepared safety seminars called "Safety Talks!" on CD, as well as a safety industry magazine, a wide range of subject- and jurisdiction-specific newsletters, and a line of safety program support documents (posters, clip art, T-shirts, pocket safety cards, and so on). Visit Safety Smart online at www.safetysmart.com.

APPENDIX H
Workplace-Specific Hazard Training Confirmation Form

After assessing the workplace to determine certified members' needs for training in how to handle workplace-specific hazards, employers must complete the following form. One copy of the form must be submitted to the WSIB, and another copy must be retained for reference within the workplace.

WSIB CSPAAT ONTARIO

Workplace-Specific Hazard Training Confirmation To WSIB (Part Two)

COMPLETE ONE FORM FOR EACH EMPLOYEE

Please complete all information. Type or print clearly

| WSIB Firm Number | WSIB Rate Group |

Employer Information
- Employer Name
- Address (including Unit number if applicable)
- City | Province | Postal Code | Telephone | Fax
- Employer Health & Safety Contact | Contact E-mail Address
- Employer Health & Safety Contact's Signature | Contact's Telephone Number | Date (dd/mmm/yyyy)

Employee Information
- Employee Name | WSIB Certification ID Number (required)
- Address (including Unit number if applicable) | Work E-mail Address
- City | Province | Postal Code | Home E-mail Address (optional)
- Employee's Signature | Telephone Number | Date (dd/mmm/yyyy)

Confidentiality Note
Your personal information is collected under the authority of the Workplace Safety and Insurance Act and will be used, disclosed and retained to administer the Certification Training program of the Workplace Safety and Insurance Board (WSIB) pursuant to the Workplace Safety and Insurance Act. Please call the WSIB at 1-800-663-6639 if you have any questions regarding the collection and use of this information.

Has a workplace hazard assessment been conducted to determine hazard topics and shared with the workplace JHSC? ☐ Yes ☐ No

Name of Hazard Topic(s) or Name of the Sector Program Completed	Date(s) Training Received (dd/mmm/yyyy)	Name of Training Provider

(see over for additional space)

Completed Form
Mail To:
Workplace Safety and Insurance Board
Certification - Prevention Services Branch
200 Front Street West, 11th Floor
Toronto ON M5V 3J1

or Fax To:
(416) 344-4921

3189A (03/06)   For more information call 1-800-663-6639 or visit our web site at www.wsib.on.ca   (1 of 2)

Workplace-Specific Hazard Training Confirmation To WSIB (Part Two)

Please complete all information.
Type or print clearly

Employee Name	WSIB Cerification ID Number (required)

Name of Hazard Topic(s) or Name of the Sector Program Completed	Date(s) Training Received (dd/mmm/yyyy)	Name of Training Provider

3189A2 (03/06)   For more information call 1-800-663-6639 or visit our web site at www.wsib.on.ca

Source: Workplace Safety and Insurance Board, "Workplace-Specific Hazard Training Confirmation to WSIB" of form # 3189A, http://www.wsib.on.ca/wsib/wsibsite.nsf/public/FormsEmployers.

APPENDIX I

WSIB Form 0007A—Employer's Report of Injury/Disease

WSIB ONTARIO CSPAAT

Mail To:
200 Front Street West
Toronto ON M5V 3J1

OR Fax To:
416-344-4684
OR 1-888-313-7373

Please PRINT in black ink

7 Employer's Report of Injury/Disease (Form 7)

Claim Number

A. Worker Information

Job Title/Occupation (at the time of accident/illness - do not use abbreviations)

Length of time in this position while working for you

Social Insurance Number

Please check **if** this worker is a: ☐ executive ☐ elected official ☐ owner ☐ spouse or relative of the employer

Is the worker covered by a Union/Collective Agreement? ☐ yes ☐ no

Worker Reference Number

Last Name

First Name

Address (number, street, apt., suite, unit)

City/Town   Province   Postal Code

Worker's preferred language ☐ English ☐ French ☐ Other

Date of Birth   dd mm yy

Telephone ( )

Sex ☐ M ☐ F

Date of Hire   dd mm yy

B. Employer Information

Trade and Legal Name (if different provide both)

Check one: ☐ Firm Number OR ☐ Account Number

Provide Number

Mailing Address

Rate Group Number

Classification Unit Code

City/Town   Province   Postal Code

Telephone ( )

Description of Business Activity

Does your firm have 20 or more workers? ☐ yes ☐ no

FAX Number ( )

Branch Address where worker is based (if different from mailing address - no abbreviations)

City/Town   Province   Postal Code

Alternate Telephone ( )

C. Accident/Illness Dates and Details

1. Date and hour of accident/Awareness of illness   dd mm yy   ☐ AM ☐ PM

Date and hour reported to employer   dd mm yy   ☐ AM ☐ PM

2. Who was the accident/illness reported to? (Name & Position)

Telephone ( )   Ext.

3. Was the accident/illness:
☐ Sudden Specific Event/Occurrence
☐ Gradually Occurring Over Time
☐ Occupational Disease
☐ Fatality

4. Type of accident/illness: **(Please check all that apply)**
☐ Struck/Caught   ☐ Fall   ☐ Slip/Trip
☐ Overexertion   ☐ Harmful Substances/Environmental   ☐ Motor Vehicle Incident
☐ Repetition   ☐ Assault
☐ Fire/Explosion   ☐ Other

5. Area of Injury (Body Part) - **(Please check all that apply)**

		Left	Right	Left	Right	Left	Right	Left	Right
☐ Head ☐ Face ☐ Eye(s) ☐ Ear(s)	☐ Teeth ☐ Neck ☐ Chest	☐ Upper back ☐ Lower back ☐ Abdomen ☐ Pelvis		Shoulder Arm Elbow Forearm		Wrist Hand Finger(s)		Hip Thigh Knee Lower Leg		Ankle Foot Toe(s)	

☐ Other

6. Describe what happened to cause the accident/illness and what the worker was doing at the time (lifting a 50 lb. box, slipped on wet floor, repetitive movements, etc...). Include what the injury is and any details of equipment, materials, environmental conditions (work area, temperature, noise, chemical, gas, fumes, other person) that may have contributed. **For a condition that occurred gradually over time, please attach a description of the physical activity required to do the work.**

A guide to complete this form is available at www.wsib.on.ca

0007A (11/05)   Page 1 of 4

WSIB / CSPAAT

7 Employer's Report of Injury/Disease (Form 7)

Claim Number

Please PRINT in black ink

Worker Name

Social Insurance Number

C. Accident/Illness Dates and Details (Continued)

7. Did the accident/illness happen on the employer's premises (owned, leased or maintained)? ☐ yes ☐ no
Specify where (shop floor, warehouse, client/customer site, parking lot, etc..).

8. Did the accident/illness happen outside the Province of Ontario? ☐ yes ☐ no
If **yes,** where (city, province/state, country).

9. Are you aware of any witnesses or other employees involved in this accident/illness? ☐ yes ☐ no
If **yes,** provide name(s), position(s), and work phone number(s).
1.
2.

10. Was any individual, who does not work for your firm, partially or totally responsible for this accident/illness? ☐ yes ☐ no
If **yes,** please provide name and work phone number

11. Are you aware of any prior similar or related problem, injury or condition? ☐ yes ☐ no
If **yes,** please explain

12. If you have concerns about this claim, attach a written submission to this form. ☐ submission attached

D. Health Care

1. Did the worker receive health care for this injury? ☐ yes ☐ no   If **yes,** when: dd mm yy
2. When did the employer learn that the worker received health care? dd mm yy

3. Where was the worker treated for this injury? **(Please check all that apply)**
☐ On-site health care ☐ Ambulance ☐ Emergency department ☐ Admitted to hospital ☐ Health professional office ☐ Clinic
☐ Other:

Name, address and phone number of health professional or facility who treated this worker (if known)

E. Lost Time - No Lost Time

1. Please choose one of the following indicators.   After the day of accident/awareness of illness, this worker:
☐ Returned to his/her **regular job** and **has not** lost any time and/or earnings. **(Complete sections G and J).**
☐ Returned to **modified work** and **has not** lost any time and/or earnings. **(Complete sections F, G, and J).**
☐ **Has** lost time and/or earnings. **(Complete ALL remaining sections).**

▶ Provide date worker first lost time   dd mm yy
▶ Date worker returned to work (if known)   dd mm yy
☐ regular work
☐ modified work

2. This Lost Time - No Lost Time - Modified Work information was confirmed by:
☐ Myself   ☐ Other   Name _____   Telephone ( )   Ext.

F. Return To Work

1. Have you been provided with work limitations for this worker's injury? ☐ yes ☐ no
2. Has modified work been discussed with this worker? ☐ yes ☐ no
3. Has modified work been offered to this worker? ☐ yes ☐ no
If **yes,** was it ☐ Accepted ☐ Declined
☐ If Declined please attach a copy of the written offer given to the worker.

4. Who is responsible for arranging worker's return to work
☐ Myself   ☐ Other   Name _____   Telephone ( )   Ext.

0007A (11/05)

Employer's Report of Injury/Disease (Form 7)

WSIB ONTARIO / CSPAAT

Please PRINT in black ink

Claim Number: _____

Worker Name: _____

Social Insurance Number: _____

G. Base Wage/Employment Information - (Do not include overtime here)

1. Is this worker (Please check all that apply)
- ☐ Permanent Full Time
- ☐ Permanent Part Time
- ☐ Temporary Full Time
- ☐ Temporary Part Time
- ☐ Casual/Irregular
- ☐ Seasonal
- ☐ Contract
- ☐ Student
- ☐ Unpaid/Trainee
- ☐ Other
- ☐ Registered Apprentice Optional Insurance
- ☐ Owner Operator or (Sub) Contractor

2. Regular rate of pay $ _____ per ☐ hour ☐ day ☐ week ☐ other

H. Additional Wage Information

1. Net Claim Code or Amount — Federal: _____ Provincial: _____

2. Vacation pay - on each cheque? ☐ yes ☐ no — Provide percentage: _____ %

3. Date and hour last worked — dd ___ mm ___ yy ___ ☐ AM ☐ PM

4. Normal working hours on last day worked — From ___ ☐ AM ☐ PM To ___ ☐ AM ☐ PM

5. Actual earnings for last day worked $ _____

6. Normal earnings for last day worked $ _____

7. Advances on wages: Is the worker being paid while he/she recovers? ☐ yes ☐ no — If yes, indicate: ☐ Full/Regular ☐ Other

8. Other Earnings (Not Regular Wages): Provide the **total of additional earnings** for each week for the 4 weeks before the accident/illness.

*For Rotational Shift workers - If the shift cycle exceeds 4 weeks, please attach the earnings information for the last complete shift cycle prior to the date of accident/illness.

▼ Use these spaces for any other earnings (indicate Commission, Differentials, Premiums, Bonus, Tips, In Lieu %, etc..).

Period	From Date (dd/mm/yy)	To Date (dd/mm/yy)	Mandatory Overtime Pay	Voluntary Overtime Pay	Commission	Commission	Commission	Commission
Week 1			$	$	$	$	$	$
Week 2			$	$	$	$	$	$
Week 3			$	$	$	$	$	$
Week 4			$	$	$	$	$	$

I. Work Schedule (Complete either A, B or C. Do not include overtime shifts)

☐ **(A.) Regular Schedule** - Indicate normal work days and hours.

Sunday	Monday	Tuesday	Wednesday	Thursday	Friday	Saturday

▶ **Example:** Monday to Friday, 40 hours

S	M	T	W	T	F	S
	8	8	8	8	8	

or,

☐ **(B.) Repeating Rotational Shift Worker** - Provide

NUMBER OF DAYS ON	NUMBER OF DAYS OFF	HOURS PER SHIFT(s)	NUMBER OF WEEKS IN CYCLE

▶ **Example:** 4 days on, 4 days off, 12 hours per shift, 8 weeks in cycle.

or,

☐ **(C.) Varied or Irregular Work Schedule** - Provide the total number of regular hours and shifts for each week for the 4 weeks prior to the accident/illness. (Do not include overtime hours or shifts here).

	Week 1	Week 2	Week 3	Week 4
From/To Dates (dd/mm/yy)	/	/	/	/
Total Hours Worked				
Total Shifts Worked				

J.

It is an offence to deliberately make false statements to the Workplace Safety and Insurance Board. I declare that all of the information provided on pages 1, 2, and 3 is true.

Name of person completing this report (please print): _____ Official title: _____

Signature: _____ Telephone: ( ) _____ Ext.: _____ Date: dd ___ mm ___ yy ___

THE WORKPLACE SAFETY AND INSURANCE ACT REQUIRES YOU GIVE A COPY OF THIS FORM TO YOUR WORKER

0007A (11/05)

Employer's Report of Injury/Disease (Form 7)

Please PRINT in black ink

Worker Name

Claim Number

Social Insurance Number

K. Additional Information

THE WORKPLACE SAFETY AND INSURANCE ACT REQUIRES YOU GIVE A COPY OF THIS FORM TO YOUR WORKER

Source: Workplace Safety and Insurance Board, "Employer's Report of Injury/Disease," form 0007A, http://www.wsib.on.ca/wsib/wsibsite.nsf/public/FormsEmployers.

APPENDIX J
Personal Perceptions of Organizational Safety

Safety Performance Solutions Inc.

Statement	Highly agree	Agree	Not sure	Disagree	Strongly disagree
1. The risk level of my job concerns me quite a bit.	1	2	3	4	5
2. When told about safety hazards, supervisors are appreciative and try to correct them quickly.	1	2	3	4	5
3. My immediate supervisor is well-informed about relevant safety issues.	1	2	3	4	5
4. It is the responsibility of each employee to seek out opportunities to prevent injury.	1	2	3	4	5
5. At my plant, work productivity and quality usually have a higher priority than work safety.	1	2	3	4	5
6. The managers in my plant really care about safety and try to reduce risk levels as much as possible.	1	2	3	4	5
7. When I see a potential safety hazard (e.g., oil spill), I am willing to correct it myself if possible.	1	2	3	4	5
8. Management places most of the blame for an accident on the injured employee.	1	2	3	4	5
9. "Near misses" are consistently reported and investigated at our plant.	1	2	3	4	5
10. I am willing to warn my co-workers about working unsafely.	1	2	3	4	5
11. Employees seen behaving unsafely in my department are usually given corrective feedback by their coworkers.	1	2	3	4	5
12. Compared to other plants, I think mine is rather risky.	1	2	3	4	5

Statement	Highly agree	Agree	Not sure	Disagree	Strongly disagree
13. Working safely is the Number One priority in my plant.	1	2	3	4	5
14. I have received adequate job safety training.	1	2	3	4	5
15. Many first-aid cases in my plant go unreported.	1	2	3	4	5
16. Information needed to work safely is made available to all employees.	1	2	3	4	5
17. Management here seems genuinely interested in reducing injury rates.	1	2	3	4	5
18. Safety audits are conducted regularly in my department to check the use of personal protective equipment.	1	2	3	4	5
19. I know how to do my job safely.	1	2	3	4	5
20. Most employees in my group would not feel comfortable if their work practices were observed and recorded by a co-worker.	1	2	3	4	5

Source: From E. Scott Geller, *Working Safe: How to Help People Actively Care for Health and Safety* (Radnor, PA: Chilton Book Company, 1996).

References

Adams, M., Burton, J., Butcher, F., Graham, S., McLeod, A., Rajan, R., & Whatman, R. (Department of Labour); Bridge, M. (ACC); Hill, R., & Johri, R. (Centre for Research on Work, Education and Business). (2002). *Aftermath: The social and economic consequences of workplace injury and illness.* Wellington, NZ: New Zealand Department of Labour.

Alberta Farm Injuries Report. (2006). Report compiled by Farm Accident Monitoring System (FAMS). Available from the website of the Alberta Department of Agriculture and Rural Development. http://www1.agric.gov.ab.ca.

Apprenticeship and Certification Act, 1998. (1998). SO 1998, c. 22.

Australian Industry Commission. (1995). *Work, health, and safety: Inquiry into occupational safety and health,* Report no. 47, Vols. I and II. Canberra. http://www.pc.gov.au/ic/inquiry/47workhe/finalreport/index.html.

Bill C-45, *An act to amend the Criminal Code (criminal liability of corporations).* Royal assent November 7, 2003.

Blanco, H., & Lewko, J., et al. (2001). *Managing the health and safety interests of young workers in small business.* Study sponsored by the Workplace Safety and Insurance Board and Laurentian University.

Building Code (Ontario). O. Reg. 403/97.

Calvert Group, Ltd. (2008). *Issue brief: Labor and workplace safety.* http://www.calvert.com/sri_IBLaborWorkplaceSafety.html.

Canadian Human Rights Act. (1985). RSC 1985, c. H-6.

Canadian Human Rights Commission. (2007). *A guide for managing the return to work.* http://www.chrc-ccdp.ca/publications/gmrw_ggrt/toc_tdm-en.asp.

Canadian Manufacturers & Exporters (CME)—Ontario Division, and Workplace Safety and Insurance Board (WSIB). (2001). Business results through health and safety. http://www.wsib.on.ca/wsib/wsibsite.nsf/public/BusinessResultsHealthSafety.

CBC News. (2006, April 29). Dying for a job. CBC In-depth report. http://www.cbc.ca/news/background/workplace-safety/dyingforajob.html.

CBC News. (2007, January 15). Out of sync with today's changing workplace. CBC In-depth report. http://www.cbc.ca/news/background/workplace-safety/outofsync.html.

Constitution Act, 1867. (1867). (UK), 30 & 31 Vict., c. 3.

Controlled Products Regulations, SOR/88-66.

Criminal Code. (1985). RSC 1985, c. C-46.

Critical Injury—Defined. (1990). RRO 1990, reg. 834.

Canadian Standards Association, *CSA-Z1000: Occupational health and safety management*, 2006. Available (for a fee) from the website of the Canadian Standards Association. http://www.csa.ca.

Deal, T.E. (1993). The culture of schools. In M. Shaskin & H.J. Walberg (Eds.). *Educational leadership and school cultures I.* Berkeley, CA: McCutchan, pp. 3–18.

Decision No. 954/06. (2006). 2006 ONWSIAT 1591.

Decision No. 1726/06. (2006). 2006 ONWSIAT 2576.

Edwards, C.A. (2004). You've had an accident! Managing the consequences in the post C-45 world: The accident response plan. http://www.sbhlawyers.com/media/OHSDD-Aug-10-2004.pdf.

Elgie, R.D. (1994). Current problems and future directions. Speech at the Toronto convention of the Association of Workers' Compensation Boards of Canada. http://www.wcb.ns.ca/WCBN/index_e.aspx?articleID=155@criteria=elgie.

Fire Code (Ontario). (1997). O. Reg. 388/97.

Geller, E.S. (1996). *Working safe: How to help people actively care for health and safety.* Radnor, PA: Chilton Book Company.

Hansen, L. (1994). Rate your B.O.S.S.—Benchmarking organizational safety strategy. *Professional Safety, 39*(6), pp. 37–43.

Hazardous Products Act. (1985). RSC 1985, c. H-3.

Hill, C.W.L., & Jones, G.R. (2001). *Strategic management,* 5th ed. Boston: Houghton Mifflin.

Human Rights Code (Ontario). (1990). RSO 1990, c. H.19.

Inco Ltd., R v. (2001). 54 OR (3d) 495 (CA).

Industrial Establishments Regulation. (1990). RRO 1990, reg. 851.

Innovative Fall Protection Inc. (2003). *Fall facts.* Calgary, AB. http://www.innovativefallprotection.com.

Knight, J., McCreadie, B., & Winch, J. (2004). *HR manager's guide to health and safety (Ontario ed.).* Toronto: Thomson Carswell.

Krause, T.R. (2005). *Leading with safety.* New York: John Wiley & Sons.

Makrides, L. (2004, February). The case for workplace health promotion. *Newsbeat* [Newsletter of the Canadian Association of Cardiac Rehabilitation (CACR)], *12*(1), pp. 1–4.

McKinnon v. Ontario (Correctional Services). (2005). 2005 HRTO 15, 52 CHRR 387 at para. 89.

McKinnon v. Ontario (Correctional Services). (2007). 2007 HRTO 4.

McNeill, K., & Ratelband, B. (2007, December 3). Health & safety: Protection of internal investigation reports through the assertion of solicitor–client and/or litigation privilege. http://www.mccarthy.ca/article_detail.aspx?id=3794.

Minto Developments, Inc. v. Duquette. (2005). 2005 CanLII 17899 (OLRB).

Nova Scotia (Workers' Compensation Board) v. Martin; Nova Scotia (Workers' Compensation Board) v. Laseur. (2003). [2003] 2 SCR 504, 2003 SCC 54.

Occupational Health and Safety Act. (1990). RSO 1990, c. O.1.

Ontarians with Disabilities Act, 2001. (2001). SO 2001, c. 32.

Ontario Ministry of Labour. (2002a). A guide to the occupational health and safety act.

Ontario Ministry of Labour, (2002b). A guide for joint health and safety committees (JHSCs) and representatives in the workplace.

Ontario Ministry of Labour. (2007, April 18). Workplace safety strategy preventing injuries and saving money. Media release.

Ontario Public Service Employees Union (OPSEU). (2001, May 24). The killer at work: Bill 57 would gut Health and Safety Act, attack other protections. Media release. http://www.opseu.org/news/axfax98/axfax%5F20010523b.htm.

Ontario v. Ontario Public Service Employees Union. (2005). 2005 CanLII 16183 (OLRB).

Orlitzky, M., Schmidt, F.L., & Rynes, S.L. (2003). *Corporate social and financial performance: A meta-analysis.* London: Sage Publications.

Patankar, M.S., Brown, J.P., & Treadwell, M.D. (2005). *Safety ethics: Cases from aviation, healthcare and occupational and environmental health.* Aldershot, UK: Ashgate.

Pater, R. (2005, December 12). The safety catalyst: Handling stubborn safety problems. http://www.maintenanceworld.com/Articles/paterR/Safety-Catalyst.htm.

Pattenden, T. (2001, April 26). Health & safety solutions seminar.

Provincial Offences Act. (1990). RSO 1990, c. P.33.

Rock, N., & Campbell, F. (2002). *Occupational health and safety in Ontario health care.* Markham, ON: LexisNexis Butterworths Canada, Inc.

Rosen, M.A. (2004). Engineering health and safety module and case studies. Toronto: Minerva. http://www.safetymanagementeducation.com/en/data/files/download/Documents/engineering%20hs%20module.pdf.

Roughton, J.E., & Mercurio, J.J. (2002). *Developing an effective safety culture: A leadership approach.* Woburn, MA: Butterworth-Heinemann.

Smith, G. (2000). *Work rage: Identify the problems, implement the solutions.* Toronto: Harper Collins.

Technical Standards and Safety Act, 2000. (2000). SO 2000, c. 16.

Technical Standards and Safety Authority. (2007, January 11). Media release.

Trades Qualification and Apprenticeship Act. (1990). RSO 1990, c. T.17.

Transportation of Dangerous Goods Act, 1992. (1992). SC 1992, c. 34.

United Food and Commercial Workers (UFCW) Canada Union. (2006, June 30). Media release.

United States Department of Labor, Bureau of Labor Statistics. (1999, December 16). News release.

United States Department of Labor, Bureau of Labor Statistics. (2004). Census of Fatal Occupational Injuries: Table 4.

Wilde, G.J.S. (1994). *Target risk.* Toronto: PDE Publications.

Workers Health and Safety Centre (Ontario). (2003, November). *From rhetoric to reality: Creating hazard-free workplaces in Ontario.* Discussion paper presented to Strategic Planning and Consultation Session.

Workplace Hazardous Materials Information System (WHMIS). (1990). RRO 1990, reg. 860.

Workplace Safety and Insurance Act, 1997. (1997). SO 1997, c. 16, Sch. A.

Workplace Safety and Insurance Board (WSIB). (2004, January). *Injured at work? We're here to help: A guide to the workplace safety and insurance system.* Toronto: WSIB.

Zaluski v. Valthane, Inc. (2000). 2000 CanLII 3201 (OLRB).

Glossary

appeal
the referral of a case to a higher court for a reconsideration of the decision of a lower court

audit
in the context of the WSIA, a compulsory review by Workplace Safety and Insurance Board investigators of an employer's compliance or lack of compliance with the WSIA

benefits
money to which a person is entitled from a pension plan, government support program, etc.

Canada Labour Code
a federal statute that deals with labour and employment issues, including occupational health and safety, that affect employees of the federal government or of certain federal works

case law
the law as established by the outcome of former cases

claim
an application for compensation under the terms of an insurance policy

class action
a legal action brought on behalf of the members of a group with a common grievance or interest

collective agreement
a complex contract between a union and an employer that sets out most of the terms of employment that govern unionized employees

common law
the body of law derived from custom and past court decisions rather than statutes

Criminal Code
the statute that describes the legislative component of Canada's criminal law

defence
an argument that the responding party, in a lawsuit, makes to counter the suing party's allegation of fault

due diligence
a possible defence against charges whereby a defendant asserts that it took all reasonable steps to avoid the incident that provoked the charge

duty of care
in negligence law, an obligation on the part of one person to take into account the effect of his or her actions on another person

essential service
a service formally designated, by a government authority (a municipality, province, or the government of Canada) as an essential service

fraud
the act of obtaining money or some other benefit illegally through deliberate deception

grievance
a formal allegation that something is unjust, inequitable, or illegal in unionized workplaces

Health and Safety Representative
in small workplaces, an individual who is assigned enhanced health and safety rights and responsibilities to be exercised on behalf of workers

hearing
the listening to evidence and pleadings in a court or other officially constituted body

inspection
in the OHS context, a visit to the workplace by a MOL inspector to assess compliance with the OHSA requirements

Internal Responsibility System (IRS)
a scheme, created under the OHSA, that assigns health and safety responsibilities to each workplace party

investigation
regardless of the word used to describe the visit by the MOL, any "inspection" triggered by an OHS event (accident, spill, claim, etc.) is an investigation; inquiries by the MOL into potential fraud or other employer offences are also investigations

Joint Health and Safety Committee (JHSC)
in medium and large workplaces, a team of individuals from management and workers, who are assigned enhanced health and safety rights and responsibilities

jurisdiction
the subject and/or geographical area over which the legal authority of a particular statute or court extends

jurisprudence
the collective of legal decisions

lawsuit
a formal request, by one party, for the court's assistance in obtaining, through its processes, compensation or relief as against another party

legislated technical standard
a technical standard that is imposed and/or enforced by legislation or regulations

legislation
a law passed by a parliament and codified in writing

litigation privilege
a legal concept that allows a party in a legal action to withold or refuse to disclose documents prepared "in contemplation of litigation"—that is, in preparation for the legal proceedings

litigation
legal action, usually with a court component—for example, a lawsuit

Material Safety Data Sheet (MSDS)
a document made available in the workplace that is designed to describe, to workers, the risks associated with a hazardous material and the appropriate safety precautions that should be followed in handling the material

Ministry of Labour (MOL)
a ministry of the Ontario government that regulates many facets—including occupational health and safety—of labour and employment in the province

negligence
the failure of a person to respect or carry out a duty of care owed to another

no-fault
(of insurance, etc.) a system under which compensation is available for losses without the need to formally attribute fault for those losses to any particular party

occupational disease
in broad terms, any disease that ensues after a workplace exposure and that has been demonstrated to be associated with that type of exposure

Occupational Health and Safety Act (OHSA)
an Ontario Statute that regulates the promotion of worker health and safety in most Ontario workplaces

occupational injury
any injury that occurs while working, or while at one's workplace

Ontario Labour Relations Board (OLRB)
an independent tribunal that mediates, arbitrates, or adjudicates certain disputes relating to labour and employment in Ontario

plea bargaining
an arrangement between a prosecutor and defendant whereby the defendant pleads guilty to a less serious charge with the expectation of leniency rather than be tried for a more serious one

premium
an amount of money paid for an insurance policy, usually at regular intervals

prosecution
a trial of a person in a court of law for a criminal offence

regulation
a legal instrument that is subservient to a statute and created to provide guidance for the administration of the statute

reprisal
in the context of occupational health and safety, any reprimand, penalty, or negative consequence imposed on an employee by an employer in response to the employee's excercise of OHS rights

risk adversity
the degree to which a person is motivated to avoid risk

risk intolerance
the degree to which a person is unwilling to accept risk

solicitor–client privilege
a legal concept that permits a party in a legal action (including a party facing prosecution under a statute) to withold or refuse to disclose documents generated between that party and his or her legal counsel

statute
a written law passed by a parliament

summary conviction offence
a minor offence that is tried using simplified criminal procedures

technical standard
an industry-recognized specification or set of specifications relating to a material, product, or service

tort
a wrongful act for which damages can be sought in a civil court by the injured party

union
an association of workers formed to protect and advance their rights and interests

voluntary technical standard
a technical standard that is published or circulated within an industry as a suggested guidleline

warrant
an order, issued by a court or court officer (such as a justice of the peace), permitting an activity (such as an entry and search) that might otherwise be refused by the party against which it is exercised

workers' compensation
money paid to a person for an injury suffered on the job

Workplace Hazardous Materials Information System (WHMIS)
a coordinated system, supported by legislation at both the federal and provincial levels, for the safe management of hazardous materials in workplaces

Workplace Safety and Insurance Act (WSIA)
the statute that governs the application of the Ontario government's scheme for the compensation of workers injured or sickened on the job

Workplace Safety and Insurance Board (WSIB)
an Ontario tribunal charged with the mediation or adjudication of disputes arising under employment-related legislation, like the OHSA

Index

accident
    defined, WSIA, 77
    without injuries, reporting requirements, 171
    zero tolerance for, 135
agricultural workers
    level of risk, 11
    protection of, 3
appeal, defined, 7
Apprenticeship and Certification Act, 1998, 103
audit, WSIA, 85

benefits, defined, 4
building codes, 100–101

CAD-7, 84, 131
Canada Labour Code
    joint management–employee health and safety committees, 18
    passage of, 17, 18
Canadian Agricultural Injury Surveillance Program, 11
Canadian Centre for Occupational Health and Safety, 14
Canadian Environmental Protection Act, 20
Canadian Human Rights Act, 108
case law, 38
claim, 1
class action, 120
codes of practice, 27
collective agreement, 41
common law, defined, 12
compensable injury, WSIA, 76
Compulsory Automobile Insurance Act, 23
confidentiality, personal information, 49
Constitution Act, 1867, 16
contributory negligence, 13
controlled products, use guidelines, 243–245
coroners' inquests
    employer's role, 207
    purpose of, 206
    triggering event, 206–207
Criminal Code
    amendments, criminal negligence, 16, 116–117
    prosecutions under, 204–206
    workplace violence, 114–116

criminal negligence, 16, 116–117
CSA-Z1000, 106, 140–142
culture of safety, 139, 228–237

Dangerous Goods Transportation Act, 103
defence, defined, 13
Drug and Pharmacies Regulation Act, 20
dry-cleaning workers, 20–21
due diligence, 200
duty of care
    architects and engineers, 35
    corporate officers and directors, 34
    defined, 34
    employers, 35–38
    licensees, 35
    owners, 34
    supervisors, 38–39
    suppliers, 35

early and safe return to work, *see* return to work (WSIA)
employee empowerment councils, 43–44
employer fraud, WSIA, 88–89
employer responsibilities, *see also* return to work (WSIA)
    duty of care, 35
    facilities and equipment maintenance, 36
    human resources management and training, 35–36
    occupational health and safety program, 36
    post-incident compliance, 36–38
    under WSIA, 78–79
enforcement, OHSA, 30–32
Environmental Protection Act, 20
essential service, 185

Fair Practices Commission, 224
Family Law Act, 23
farm workers
    level of risk, 11
    protection of, 3
fellow-worker rule, 13
fire codes, 101
flu pandemic, 152–153
fraud, 85

grievance, 170, 203–204

hazard complaints
    external, 168–169
    internal, 167
Hazardous Materials Information Review Act, 65
Hazardous Materials Information Review Commission, 65
health and safety month, 158
Health and Safety Representative, 26, 39, 44–45
health promotion programs, 155
health screen, 110
hearing, WSIAT, 82
Hogg's Hollow accident, 15
Human Rights Code, 108
human rights legislation
    co-worker harassment, 112
    disability-related safety issues, 109
    discrimination in hiring, 109
    health screening, 110
    immunization, 111–112
    incidents concerning, 164
    Ontarians with Disabilities Act, 113
    safety, and, 108
    substance-use testing, 111

immunization, 111–112
incidents
    accidents without injuries, 171
    complaints about hazards/conditions, 166
        external complaints, 168–169
        internal complaints, 167
    costing form, 255–257
    critical injuries and fatalities, 177–178
    defined, 162
    environmental incidents, 164
    examples of, 163–164
    first aid/no-lost-time injuries, 171
    human rights incidents, 164
    incidents requiring action, 166
    lost-time/medical-attention injuries, 172
        immediate considerations, 172–173
        reporting obligations, 173–176
    work refusals/stoppages, 169–170
Industrial Accident Prevention Association, 97
Industrial Establishments regulation
    application, 94
    Pre-Start Health and Safety Review, 96–97
    safety specifications under, 95
industrial hygiene, 9
industrial safety standards, 105–106
industry-specific safety statutes, 106–107

injury cost calculator, 6
inspections
    defined, 188
    Ministry of Labour
        appeals from, 191–193
        employer obligations, 189–190
        inspectors' orders, 190
        legal representation, 193–194
        reasons for, 188–189
Internal Responsibility System (IRS)
    defined, 24
    employer role, 35–38
    Health and Safety Representative, 39, 44–45
    Joint Health and Safety Committee, 39, 45–50
    key participants, 34
    supervisor role, 38–39
    trade unions, and, 39–42
    worker role, 42–44
investigation
    defined, 194
    differentiated from inspection, 194–196
    employer obligations, 196–198
    follow-up actions, 183
    investigation team, formation of, 182–183
    management of, 157
    objectives, 183
    serious accidents, 49–50

Joint Health and Safety Committee (JHSC)
    certification of member, 47
    composition of committee, 45–46
    defined, 26
    mandate and duties, 47–48
    multi-workplace committees, 47
    procedures
        handling worker complaints, 49
        inspections, 48
        meetings, 48
        serious accident investigations, 49–50
    role in IRS, 39
jurisdiction, 13
jurisprudence, 41

lawsuit, 19
legislated technical standards
    building codes, 100–101
    dangerous goods and materials, 102–103
    defined, 94
    examples of, 98
    fire codes, 101–102
    resources, safety in use of, 102

Technical Standards and Safety Act, 98
Technical Standards and Safety Authority, 99
training and certification standards, 103–104
legislation, 7
litigation, defined, 12
litigation privilege, 199

Maquila Solidarity Network, 120–121
Material Safety Data Sheet (MSDS)
    completion
        employers, 66–69
        sample, 247–254
        suppliers/importers, 60–65
    compliance statistics, 66
    defined, 55
mental stress, 78
Merit-Adjusted Premium Plan for Small Business, 84, 131
Mining Operations Act, 93
Ministry of Labour (MOL)
    inspections, 188–194
    investigations, 194–198
    prosecutions, 198–202
    role of, 7

National Institute of Occupational Safety and Health, 6
negligence, defined, 16
New Experimental Experience Rating (NEER) system, 83–84, 131, 139, 146
no-fault insurance, 13
notices, under OHSA, 30, 40

occupational disease
    defined, 14
    WSIA, and, 77–78
Occupational Health and Safety Act (OHSA)
    amendments, agricultural workers, 3–4
    Industrial Establishments regulation under, 94–97
    Internal Responsibility System
        employer role, 35–38
        Health and Safety Representative, 39, 44–45
        Joint Health and Safety Committee, 39, 45–50
        key participants, 34
        supervisor role, 38–39
        trade unions, and, 39–42
        worker role, 42–44
    reporting requirements, 174
    scheme-creating legislation, as, 23
    structure of
        administration, 26–27
        application, 25–26
        Codes of Practice, 27
        definitions, 24–25
        duties of employers, 27
        employer reprisals, 30
        enforcement, 30–32
        notices, 30
        offences and penalties under, 32
        regulations, 32–33, 94–97, 239–240
        right to refuse/stop work, 28–30
        table of contents, 24
        toxic substances, 28
occupational health and safety management system
    building health and safety system, 139–142
    change, planning for, 146–147
    evaluation culture, perceptions, behaviours, 145–146
    evaluation formal compliance and reporting, 146
    Health and Safety Representative, appointment, 150
    Health Promotion Programs, 155–156
    injury and accident statistics, review of, 143–144
    inspections, scheduling, 151
    investigations, management of, 157
    Joint Health and Safety Committee, formation, 150
    policies and procedures
        development of, 151–152
        review of, 144–145
    resources, marshalling, 147–149
    senior management, role of, 138–139
    training, 153–155
    WHMIS compliance, 150–151
occupational illnesses
    defined, 179
    reporting requirements, 179–180
occupational injuries, *see also* incidents
    critical injuries, 177–178
    defined, 14
    needing medical attention, 172–176
    no time lost, 171–172
Office of the Employer Adviser, 82
Ontarians with Disabilities Act, 109, 113
Ontario Human Rights Commission, 109
Ontario Labour Relations Board (OLRB), 7

penalties under OHSA, 32
Physical Demands Information Form, 211–212
plea bargaining, 205
post-traumatic stress disorder, 181–182
premiums
    calculation, 79, 80
    classification, challenges to, 81
    defined, 72
    factors influencing, 80
    good safety performance, and, 83–84

Pre-Start Health and Safety Review, 96–97
prosecution (MOL)
    bad press, minimization, 201–202
    criminal prosecutions, 204–206
    defined, 7
    factors influencing fines, 201
    preparation for court, 199–201
    privileged documents, 198–199
Provincial Offences Act, 107

regulations under OHSA, 32–33
reporting requirements, post-incident, 173–178, 265–268
reprisal
    defined, 29
    prohibition, 30
    WHMIS inquiries, 56
resources, safe use of, 102
return to work (WSIA)
    accommodation, planning for, 220–221
    assessment, date/abilities/limitations, 210–213
    co-workers, and, 221–222
    complaints, 222–223
    duty to support, expiration of, 216–217
    legal obligation to promote, 210
    post-accident communication, 218–220
    suitable work, identification of, 213
    timing and preparations, 221
    unlikely event, 224–225
    worker accommodation, 214–216
    worker's obligation, 217
risk, voluntary assumption of, 13
risk adversity, 231
risk intolerance, 233

Safe Communities Incentive Program, 132
Safe Drinking Water Act, 20
Safety Groups Program, 131
safety leadership, 134–135
Second Injury Enhancement Fund, 131, 216
social costs, 8
solicitor–client privilege, 199
Standards Council of Canada, 105
statute, defined, 12
subject-specific safety statutes, 106–107
substance-use testing, 111
summary conviction offence, 126

technical standard, 94
tort, defined, 12
Trades Qualification and Apprenticeship Act, 104

Transportation of Dangerous Goods Act, 1992, 103
traumatic events, 180–182

union
    defined, 3
    OHSA, and, 39–42, 203–204

victim fine surcharges, 107
violence in the workplace, 114–116
voluntary assumption of risk, 13
voluntary technical standards
    defined, 94
    sources of, 105–106

warrant, 31
Westray Mine disaster, 15–16
work refusals, 169–170
worker, defined, WSIA, 75
worker fraud, WSIA, 85, 87–88
worker obligations, OHS, 42–44
workers' compensation
    defined, 4
    legislation, introduction of, 13
Workmen's Compensation Act, 13, 18, 72, 77
Workmen's Compensation for Injuries Act, 13, 18
Workplace Hazardous Materials Information System (WHMIS)
    classification, hazardous materials, 57, 58
    defined, 28
    employer compliance, 66–69
    goal of, 53–54
    history of, 54
    introduction of, 53
    labels, 61, 63
    legislative framework, 55
        federal aspects, 55
        provincial aspects, 55–56
    supplier/importer compliance, 57–66
workplace safety
    approaches to, 8
    associations, 241–242
    cost–benefit analysis, 119–134
    economics of, 5–6
    employee/union perspective, 2–4
    employer/management
        perspective, 4–6
        responsibilities, 35–38
    government's perspective, 7–8
    injuries, cost of
        administrative costs, 125
        estimation form, 123

    fines and other penalties, 125–126
    lost productivity, 124
    property damage, 123–124
    public image, impact on, 127
    recruitment and training costs, 124
    society costs, 127–128
law, evolution of, 11–21
safety promotion, costs of
    categories of costs, 129–130
    OHS costs, justification, 130
    opportunities to offset, 131–133
society's perspective, 6–7

Workplace Safety and Insurance Act (WSIA)
    application, 72
        general, 72
        schedules, 72–75
    benefits, types of, 84, 86–87
    compensable injuries, 76–77
    employer duties, 78–79
    employer fraud, 88–89
    enactment, 71
    fraud, management of, 85, 87–88
    occupational diseases, 77
    offences and penalties under, 89
    premium management, 79–84

    reporting requirements, 174–176
    return to work
        accommodation, planning for, 220–221
        assessment, date/abilities/limitations, 210–213
        co-workers, and, 221–222
        complaints, 222–223
        duty to support, expiration of, 216–217
        general, 85
        legal obligation to promote, 210
        post-accident communication, 218–220
        suitable work, identification of, 213
        timing and preparations, 221
        unlikely event, 224–225
        worker accommodation, 214–216
        worker's obligation, 217
    worker, defined, 75
Workplace Safety and Insurance Appeals Tribunal (WSIAT), 75, 82
Workplace Safety and Insurance Board (WSIB)
    audit documents, 144–145
    defined, 36
    investigations and prosecutions, 202
Workplace-Specific Hazard Training Confirmation Form, 263–264
Workwell audits, 202–203